T0260803

LTspice® for Linear Circuits

LTspice® for Linear Circuits

James A. Svoboda
Professor Emeritus, Clarkson University
Potsdam, NY, USA

This edition first published 2023
© 2023 John Wiley & Sons, Inc.

Published by John Wiley & Sons, Inc., Hoboken, New Jersey.
Published simultaneously in Canada.

For general information on our other products and services or for technical support,
please contact our Customer Care Department within the United States at (800)
762-2974, outside the United States at (317) 572-3993 or fax (317) 572-4002.

Wiley also publishes its books in a variety of electronic formats. Some content that
appears in print may not be available in electronic formats. For more information about
Wiley products, visit our web site at www.wiley.com.

Library of Congress Cataloging-in-Publication Data
Names: Svoboda, James A., author.
Title: LTspice® for linear circuits / James A. Svoboda.
Description: Hoboken, New Jersey : Wiley, 2023. | Includes bibliographical
    references and index.
Identifiers: LCCN 2023007237 (print) | LCCN 2023007238 (ebook) | ISBN
    9781119987925 (cloth) | ISBN 9781119987956 (adobe pdf) | ISBN
    9781119987963 (epub)
Subjects: LCSH: Electric circuits, Linear. | Electronic circuits–Computer simulation.
Classification: LCC TK454 .S926 2023 (print) | LCC TK454 (ebook) | DDC
    621.319/20113–dc23/eng/20230224
LC record available at https://lccn.loc.gov/2023007237
LC ebook record available at https://lccn.loc.gov/2023007238

Cover Design: Wiley
Cover Image: Courtesy of James A. Svoboda

Set in 9.5/12.5pt STIXTwoText by Straive, Pondicherry, India

Contents

Preface

In their sophomore year engineering students take a course in which they learn to analyze electric circuits. Textbooks for this course have titles like *Introduction to Electric Circuits, Engineering Circuit Analysis*, and *Linear Circuits*. These textbooks have very similar tables of contents, and they all advocate analyzing electric circuits by writing and solving sets of simultaneous equations called "mesh equations" or "node equations." Writing mesh equations or node equations is relatively easy, so most of the effort required to analyze an electric circuit is in solving those simultaneous equations.

Another approach is to use LTspice to determine the values of the mesh currents and node voltages of a circuit. In this approach, we do not need to solve simultaneous equations to analyze a circuit. We do, however, need to check the values of the mesh currents and node voltages to protect against data entry errors because it is easy to make errors when entering data by hand; a simple typo will cause the output to be wrong.

We verify that the answers are correct by substituting the values of the mesh currents provided by LTspice into the mesh equations or by substituting the values of the node voltages provided by LTspice into the node equations. *LTspice for Linear Circuits* provides a detailed description of how to represent a given electric circuit as an LTspice schematic, how to display the "simulation results," and then how to verify that the results are correct.

In *LTspice for Linear Circuits*, we use a six-step procedure to organize circuit analysis, stated as follows:

Step 1. Formulate a circuit analysis problem.
Step 2. Describe the circuit using an LTspice schematic.
Step 3. Simulate the circuit using LTspice.
Step 4. Display the results of the simulation.
Step 5. Verify that the simulation results are correct.
Step 6. Report the answer to the circuit analysis problem.

The second, third, and fourth steps of this procedure actually use LTspice. In the first step, the user identifies the problem that is to be solved and in the fifth step verifies that it has indeed been solved.

It is hard to overemphasize the importance of steps 1 and 5. If the simulation is not correct, the error must be identified and eliminated so that a correct simulation can be performed. Only after verifying that the simulation is correct can the answer to the circuit analysis problem be reported. The benefit provided by using LTspice with this six-step procedure is the requirement that we check each answer obtained using that procedure.

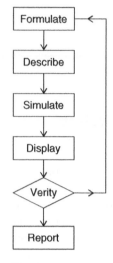

The six-step procedure for circuit analysis using LTspice.

Chapter 1

Getting Started with LTspice

1.1 Introduction

SPICE is a computer program used for numerical analysis of electric circuits. SPICE is an acronym for Simulation Program with Integrated Circuit Emphasis. Developed in the early 1970s at the University of California at Berkley, SPICE is generally regarded to be the most widely used circuit simulation program [1]. LTspice is a version of SPICE for personal computers produced by Analog Devices (originally by Linear Technology) [2].

Figure 1.1 shows the opening screen of LTspice. Figure 1.1 identifies three ways of interacting with LTspice. We issue commands to LTspice using the Window Tabs and Toolbar icons. The Status bar displays messages from LTspice. (The LTspice toolbar can be docked, as shown in Figure 1.1 or undocked as shown in Figure 1.2. Double-click on the toolbar to undock a docked toolbar or to dock an undocked toolbar.)

Figure 1.2 identifies the commands associated with some of the icons on the toolbar. Many of those same commands are available from the Window Tabs. Figure 1.3 shows the commands that are available from the "Edit" Window tab.

A circuit diagram is called a schematic in LTspice. Consider representing the circuit shown in Figure 1.4 as a schematic in LTspice. Begin by clicking on the New Schematic icon, ⟨⟩, on the toolbar. The LTspice screen will change in a couple of ways (compare Figures 1.1 and 1.5). In particular, the LTspice Workspace will now appear white because it contains a new, blank schematic. Also, additional tabs will be available in the LTspice Window Tabs.

LTspice® for Linear Circuits, First Edition. James A. Svoboda.
© 2023 John Wiley & Sons, Inc. Published 2023 by John Wiley & Sons, Inc.

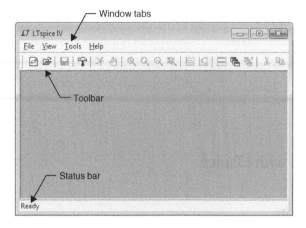

Figure 1.1 The opening screen of LTspice.

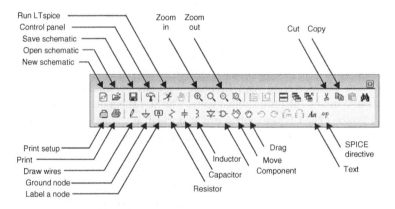

Figure 1.2 Identifying icons on the LTspice toolbar.

Drafting a schematic in LTspice requires four activities:

1) Place symbols representing the circuit elements on the schematic.
2) Adjust the values of the circuit element parameters, e.g. the resistances of the resistors, the voltage of the voltage source, and the current of the current source.
3) Place a ground symbol to identify the bottom node of the schematic as the ground node. (Notice that the bottom node of the circuit in Figure 1.4 has been identified as the ground node.)
4) Draw the wires that connect the circuit elements.

Figure 1.3   Icons on the "Edit" Window Tab.

Figure 1.4   The circuit for the first example.

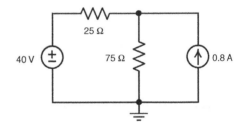

25 Ω

40 V   75 Ω   0.8 A

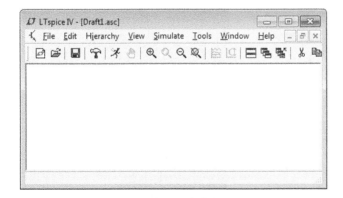

Figure 1.5   A new, blank schematic in LTspice.

The circuit shown in Figure 1.4 consists of two resistors: a voltage source and a current source. Let us begin by placing a symbol representing the voltage source in the LTspice schematic. Click on the Component icon, ⊅, on the toolbar to pop up the Select Component Symbol dialog box shown in Figure 1.6. Start typing "voltage source" in the search box. LTspice will find the symbol for a voltage source as soon as "vo" has been typed. Click the "OK" button on the Select Component Symbol dialog box. A voltage source symbol will appear on the schematic as shown in Figure 1.7. Position the voltage source symbol as desired by dragging the mouse, and then left-click to place the voltage source symbol on the schematic. (To move a symbol that was placed previously, click first on the Move toolbar icon, ✋,

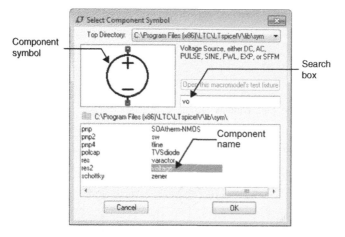

Figure 1.6 The "Select Component Symbol" dialog box.

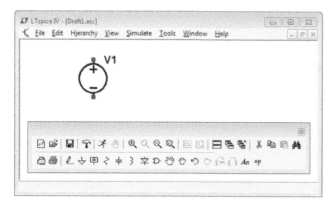

Figure 1.7 A voltage source symbol placed on the schematic.

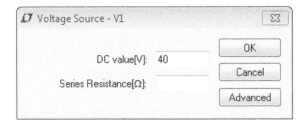

Figure 1.8 Specify the voltage of voltage source V1.

and then click the component symbol to be moved. Position the component symbol as desired by dragging the mouse, and then left-click again to place the voltage source symbol on the schematic.)

Notice that this voltage source has been labeled as element V1 in Figure 1.7. We want the voltage of V1 to be 40 V – the voltage of the voltage source in Figure 1.4. Right-click on the voltage source symbol on the schematic to pop up the dialog box shown in Figure 1.8 and set the voltage of V1 to 40 V.

Click on the resistor icon, ⤢, twice to place symbols for resistors R1 and R2 on the schematic. (The resistor icon on the toolbar provides a shortcut to clicking on the component icon, ⅅ, and then typing "resistor" in the search box in the Select Component Symbol dialog box.) Right-click on R1 and use the resulting dialog box to set the resistance of R1 to 25 Ω. Right-click on R2 and set the resistance of R2 to 75 Ω.

Click on the Component icon, ⅅ, on the toolbar to pop up the Select Component Symbol dialog box. Start typing "current source" in the search box. LTspice will find the symbol for a current source as soon as "cu" has been typed. Click the "OK" button on the Select Component Symbol dialog box. A current source symbol will appear on the schematic as shown in Figure 1.9. Position the current source symbol as desired. Right-click on the current source and set the current to 0.8 A.

Click on the Ground icon, ⏚, to place a ground symbol on the schematic.

Figure 1.9 shows the schematic after placing symbols representing the circuit elements that comprise the circuit in Figure 1.4 and adjusting the values associated with those symbols. We need to rotate the symbol of the 25 Ω resistor by 90° and to rotate the symbol of the current source by 180° to make the schematic correspond to the circuit diagram in Figure 1.4. Click on the move icon, 🖐, and then click the symbol of the 25 Ω resistor and type <Ctrl>R to rotate the symbol of the 25 Ω resistor by 90°. Position the symbol of the 25 Ω resistor as desired and left-click to finish. Similarly, Click on the move icon, 🖐, and then click the current source symbol and type <Ctrl>R twice to rotate the symbol of the current source by 180°.

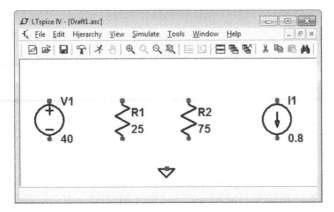

Figure 1.9 The schematic after placing symbols.

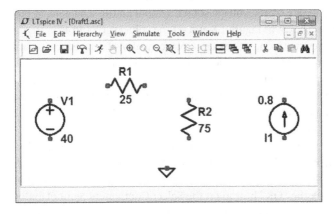

Figure 1.10 The schematic after rotating symbols.

Figure 1.10 shows the schematic after component symbols have been placed and rotated and the values of component parameters have been specified. The wires connecting these component symbols will be represented by horizontal and vertical straight lines. Notice that each component symbol identifies the locations of the terminals of the component by small squares. Consider, for example, drawing the wire connecting the top node of the voltage source symbol to the left node of the symbol of the 25 Ω resistor. This wire will be represented by two connected line segments: one vertical and one horizontal. Click on the "Draft wires" icon, ✐, on the toolbar, and then click on the top terminal of the voltage source symbol to begin drawing

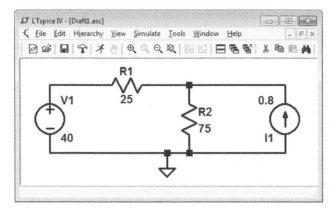

Figure 1.11 The schematic after wiring symbols.

the vertical line segment. Drag the mouse upward to draw the vertical line segment. Click to change directions, and then drag the mouse to the left terminal of the symbol of the 25 Ω resistor.

Figure 1.11 shows the circuit after adding wires to connect the circuit elements.

1.2 Six Steps

We will use a six-step procedure to organize circuit analysis using LTspice. This procedure is illustrated in Figure 1.12 and is stated as follows.

Step 1. **Formulate** a circuit analysis problem.
Step 2. **Describe** the circuit using an LTspice schematic.
Step 3. **Simulate** the circuit using LTspice.
Step 4. **Display** the results of the simulation.
Step 5. **Verify** that the simulation results are correct.
Step 6. **Report** the answer to the circuit analysis problem.

The second, third, and fourth steps of this procedure use LTspice. In the first and fifth steps, the user identifies the problem that is to be solved and verifies that it has indeed been solved. It would be hard to overemphasize the importance of steps 1 and 5.

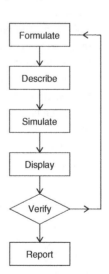

Figure 1.12 A six-step procedure for circuit analysis using LTspice.

If the simulation is not correct, the error must be identified and eliminated so that a correct simulation can be performed. Only after verifying that the simulation is correct can the answer to the circuit analysis problem be reported.

Example 1.1

This example illustrates the six-step procedure for circuit analysis from Figure 1.12 by using it to analyze the circuit shown in Figure 1.13.

Step 1. Formulate a circuit analysis problem.

Determine the value of v_2, the voltage across the 75 Ω resistor of the circuit shown in Figure 1.13. Notice that the bottom node of this circuit is identified as the reference node. Consequently, v_2 is a node voltage of the circuit.

Step 2. Describe the circuit using an LTspice schematic.

In general, the goal of circuit analysis is to determine the values of the voltages across, and currents in, the devices of the circuit. In contrast, analysis of a circuit using LTspice provides the values of the node voltages and the device currents. We can easily determine the values of the device voltages from the values of the node voltages.

Click on the New Schematic toolbar icon, 🔲, to start a new schematic. Place the component symbols, adjust the values of the circuit parameters, place a ground symbol, and wire the circuit as described on pages 1–7 to obtain the schematic shown in Figure 1.14.

Step 3. Simulate the circuit using LTspice.

LTspice is capable of performing several types of simulation. To specify the desired type of simulation, click on the Edit Tab and then select "Spice Analysis" (see Figure 1.3) to pop up the Edit Simulation Command dialog box as shown in Figure 1.15.

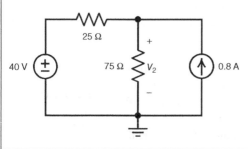

Figure 1.13 The circuit considered Example 1.1.

Example 1.1 (Continued)

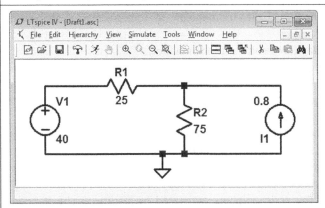

Figure 1.14 The schematic corresponding to the circuit in Figure 1.13.

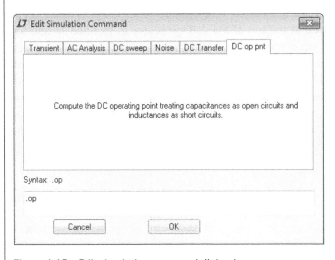

Figure 1.15 Edit simulation command dialog box.

The circuit in Figure 1.13 is a DC circuit because the values of both the voltage-source voltage and the current-source current are constants. For historical reasons SPICE refers to analyzing a DC circuit as "finding the DC operating point" of the circuit.

Select the "DC op pnt" tab as shown in Figure 1.15, and then click the "OK" button. LTspice will generate a "spice directive" consisting of the command ".op." Position the spice directive on the schematic as desired.

(Continued)

Example 1.1   (Continued)

Click on the Run LTspice toolbar icon, ⚡, to perform a DC analysis of the circuit.

Step 4. Display the results of the simulation.

The results of a DC analysis display automatically as shown in Figure 1.16. The results of the DC analysis are labeled as the "Operating Point" in Figure 1.16. The operating point consists of node voltages in Volts and the circuit element currents in Amps.

Node voltages at two nodes, labeled nodes n001 and n002, are given in Figure 1.16. We expect the node voltage at the top node of the voltage source to be 40 V. Apparently node n001 is the top node of the voltage source and node n002 is the top node of the 75 Ω resistor.

Matching the node numbers generated by LTspice to circuit nodes becomes more tedious when LTspice is used to analyze larger circuits. Instead, let us choose convenient node numbers and label the nodes of the schematic using those convenient node numbers. For example, let us label the node at the top of the voltage source to be node "a" and the node at the top of the 75 Ω resistor to be node "b." Click on the "Label a Node" toolbar icon, 🄰, (see Figure 1.2) to pop up the dialog box shown in Figure 1.17. Enter "a" in the text box and select "Output" from the drop-down menu as shown in Figure 1.17. Click the OK button to add a node label to the schematic. Position this symbol as desired and wire it to the top node of the voltage source as shown in Figure 1.18. Next, label the top node of the 75 Ω resistor to be node b as shown in Figure 1.18.

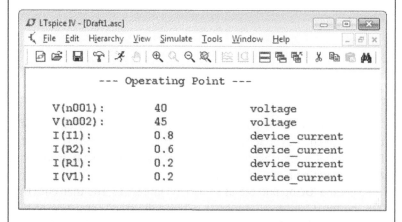

Figure 1.16   Simulation results.

Example 1.1   (Continued)

Figure 1.17   Labeling node "a."

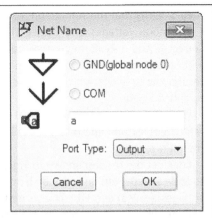

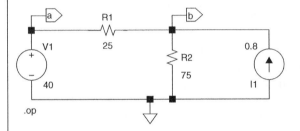

Figure 1.18   The Example 1.1 schematic after labeling nodes "a" and "b."

```
          - - - Operating Point - - -


v(a) :              40              voltage

v(b) :              45              Voltage

I(I1) :             0.8             device_current

I(R2) :             0.6             device_current

I(R1) :             0.2             device_current

I(V1) :             0.2             device_current
```

Figure 1.19   Revised simulation results.

Run the simulation again to obtain the revised simulation results shown in Figure 1.19. Notice that the names of the nodes are now "a" and "b."

(*Continued*)

Example 1.1 (Continued)

Step 5. Verify that the simulation results are correct.

The node voltages and device currents shown in Figure 1.19 are correct if, and only if, they satisfy both Ohm's and Kirchhoff's laws. We will label the circuit diagram from Figure 1.13 using the simulation results and then check if those currents and voltages satisfy Ohm's and Kirchhoff's laws.

We see from Figure 1.19 that I(R1) = 6 A. Is this the value of the current directed from right to left in the 25 Ω resistor or the current directed from left to right?

Consider the resistor shown in Figure 1.20. This resistor is connected to node a on the left and to node b on the right. The corresponding node voltages are $V(a)$ and $V(b)$. The currents i_1, directed from node a toward node b, and i_2, directed from node b toward node a, are given by

$$i_1 = \frac{V(a) - V(b)}{R} \quad \text{and} \quad i_2 = \frac{V(b) - V(a)}{R}$$

When $R > 0$ and $V(a) > V(b)$, then $i_1 > 0$ and $i_2 < 0$. The rule for assigning a direction to I(R1) is simple: when R1 > 0 and $V(a) > V(b)$ then I(R1) is the current directed from node a toward node b.

A different rule is used to assign a direction to a voltage-source current such as I(V1) in Figure 1.19. Consider the voltage source shown in Figure 1.21. The current I(V1) is directed from the node near the plus sign of the voltage polarity toward the node near the minus sign.

We can label the node voltages and device currents given in the simulation results (Figure 1.16) on the circuit diagram (Figure 1.13) as shown in Figure 1.22. We readily verify that the node voltages and device currents satisfy Ohm's and Kirchhoff's laws. (See the appendix to this chapter for an introduction to Ohm's and Kirchhoff's laws.)

Step 6. Report the result.

The value of v_2, the voltage across the 75 Ω resistor in Figure 1.13, is $v_2 = 45$ V.

Figure 1.20 Node voltages and resistor currents.

Figure 1.21 Node voltages and voltage-source currents.

Example 1.1 (Continued)

Figure 1.22 Node voltages and device currents.

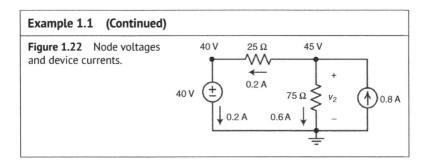

1.2.1 LTspice Notation

Figure 1.23 shows an LTspice component, in this case, a resistor. The component is labeled twice by a "Instance Name" and also by a "Value." The instance name identifies a particular component and distinguishes it from all other components in the same schematic. The instance name is a sequence of characters (letters and numbers) beginning with a letter. The initial letter corresponds to the component type as shown in Table 1.1.

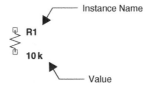

Figure 1.23 An LTspice component labeled by its "Instance Name" and its "Value."

The "Value" is a real number or an expression. In Figure 1.23 the value is "10k" where "k" is a multiplying factor abbreviation indicating 1000. That is, the value of 10k is 10 * 1000 = 10 000. Table 1.2 lists the available multiplying factors. The "Value" of a resistor represents the resistance of the

Table 1.1 LTspice components and their labels.

Component	Instance name	Value
Current source	I. . .	Source current, A
Voltage source	V. . .	Source voltage, V
Capacitor	C. . .	Capacitance, F
Inductor	L. . .	Inductance, H
Resistor	R. . .	Resistance, Ω
VCVS	E. . .	Gain, V/V
CCCS	F. . .	Gain, A/A
VCCS	G. . .	Gain, A/V
CCVS	H. . .	Gain, V/A

Table 1.2 LTspice scale factor abbreviations.

Letter suffix	Multiplying factor	Name of suffix
T	1E12	tera
G	1E9	giga
MEG	1E6	mega
K	1E3	kilo
M	1E-3	milli
MIL	25.4E-6	mil
U	1E-6	micro
N	1E-9	nano
P	1E-12	pico
F	1E-15	femto

resistor in Ω. The third column of Table 1.1 gives interpretation of the Value for several component types.

Both the Instance Name and the Value are properties of LTspice components and are specified and changed using the Component Attribute Editor. For example, to change the resistance from "10k" to "25k," <CNTL>right-click on resistor R1 to pop up the Component Attribute Editor as shown in Figure 1.24. (The Xs on the right-hand column indicate

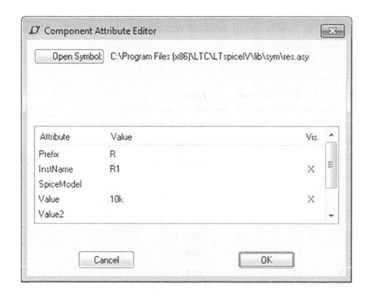

Figure 1.24 The component attribute editor.

(a) (b) (c)

R1 Rc2+ Rc2+
25k 25k {(1+4)*10k}

Figure 1.25 An LTspice component labeled by its "Instance Name" and its "Value."

that both the Instance Name and the Value of Component R1 will be visible on the schematic.) Edit the Value to change "10k" to "25k" and click the "OK" button. Component R1 will now appear as shown in Figure 1.25a.

<CNTL>Right-click on Component R1 to pop up the Component Attribute Editor again. Edit the Instance Name to change "R1" to "Rc2+" and click the "OK" button. Component R1 will now appear as shown in Figure 1.25b.

<CNTL>right-click on Component Rc2+ to pop up the Component Attribute Editor again. Edit the Value to change "25k" to "{(1 + 4) * 10k}" and click the "OK" button. Component Rc2+ will now appear as shown in Figure 1.25c. Braces indicate that the Value is an expression rather than a number. The resistance of component Rc2+ is $(1 + 4) * 10k = 50\,k\Omega$.

1.A Appendix: Verifying LTspice Simulation Results

1.A.1 Node Voltages and Device Currents

A circuit is a collection of interconnected devices. (Devices are sometimes called "circuit elements" or just "elements.") These devices are connected to each other using "leads" that have been attached to the devices for that purpose. Figure 1.A.1a shows a device that has two leads that can be connected to other devices. Figure 1.A.1b shows three devices connected to each other by their leads. The place where these leads are connected together is called a "node," as shown in Figure 1.A.1b.

Figure 1.A.2a shows a generic circuit element connected between two nodes labeled as node "a" and node "b." There are three nodes in Figure 1.A.2b: nodes a and b of the generic circuit element and a third node

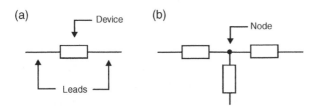

(a) Device (b) Node

Leads

Figure 1.A.1 Leads and nodes.

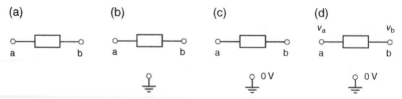

Figure 1.A.2   Nodes and node voltages.

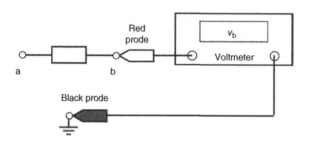

Figure 1.A.3   Measuring a node voltage in the laboratory using a voltmeter.

that has been designated as "the ground node" by attaching the ground symbol shown in Figure 1.A.2b to that node. By convention the voltage at the ground node is 0V as shown in Figure 1.A.2c. Having identified the ground node, we can associate a "node voltage" with each node of the circuit as shown in Figure 1.A.2d.

Figure 1.A.3 shows how to measure a **node voltage** using an instrument called a voltmeter. The voltmeter has two color-coded probes: a black probe and a red probe. (In a black and white drawing such as Figure 1.A.3, the red probe will be represented by a white probe.) When the black probe is connected to the ground node and the red probe is connected to node b as shown in Figure 1.A.3, the voltmeter displays the value, v_b, of the node voltage at node b. The value of v_a, the node voltage at node a, is obtained by moving the red probe from node b to node a. Often we label the nodes of a circuit with the corresponding node voltages as shown in Figure 1.A.2d.

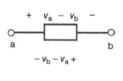

The voltage across a circuit element can be labeled in either of the two ways as shown in Figure 1.A.4. The + and − signs in these labels indicate the polarity of the element voltage. Figure 1.A.5 shows how to measure a voltage across a circuit element using a voltmeter. Notice

Figure 1.A.4   Labeling element voltages and their polarities.

(a) (b)

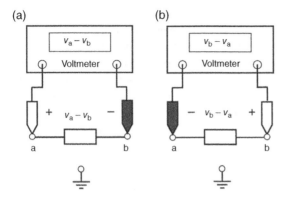

Figure 1.A.5 Measuring element voltages in the laboratory using a voltmeter.

that the locations of the color-coded probes of the voltmeter correspond to the polarity of the element voltage. Also, since

$$v_a - v_b = -(v_b - v_a) \tag{1.A.1}$$

reversing the polarity of an element voltage requires multiplying the value of that element voltage by -1.

The current in a circuit element can be labeled in either of the two ways as shown in Figure 1.A.6. Figure 1.A.7 shows how to measure a current in a circuit element using an ammeter. Reversing the direction of an element current requires multiplying that current by -1. Consequently, $i_{ab} = -i_{ba}$ in Figure 1.A.7.

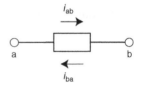

Figure 1.A.6 Labeling element currents and their directions.

(a) (b)

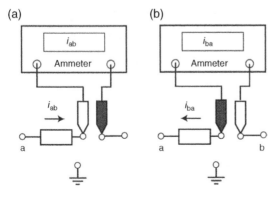

Figure 1.A.7 Measuring element currents in the laboratory using an ammeter.

The terms "passive convention" and "active convention" describe the relationship between the polarity of the element voltage of a circuit element and the direction of the element current of that circuit element. The element current and voltage are said to adhere to the **passive convention** when the element current is directed from the node near the plus sign of the element voltage polarity toward the node near the minus sign of the element voltage polarity. Conversely, the element current and element voltage are said to adhere to the **active convention** when the element current is directed from the node near the minus sign of the element voltage polarity toward the node near the plus sign of the element voltage polarity.

1.A.2   Ohm's and Kirchhoff's Laws

In 1827, Georg Simon Ohm determined that the current in a conducting wire of uniform cross section could be expressed as $i = \dfrac{Av}{\rho L}$ where A is the cross-sectional area, ρ is the resistivity, and L is the length of the wire. Also, the direction of the current, i, in the wire is required to adhere to the passive convention with the voltage, v, across the resistor. Ohm defined the resistance, R, of the wire to be $R = \dfrac{A}{\rho L}$, after which Ohm's law [3] is

$$v = Ri \qquad\qquad (1.A.2)$$

Suppose that the circuit element in Figure 1.A.8 is a resistor, as shown in Figure 1.A.9. According to Ohm's law, the voltage across a resistor is proportional to the current in the resistor. (Notice that the resistor

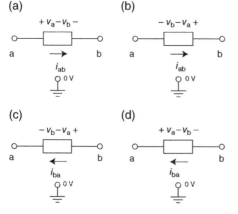

(a)

(b)

(c)

(d)

Figure 1.A.8   The element current and element voltage adhere to the passive convention in (a) and (c) and to the active convention in (b) and (d).

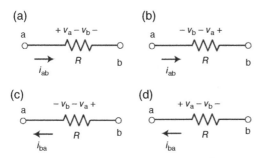

Figure 1.A.9 The resistor current and resistor voltage adhere to the passive convention in (a) and (c) and to the active convention in (b) and (d).

voltage and current labeled in Figure 1.A.8a and c adhere to the **passive convention.**) In this case

$$v_a - v_b = Ri_{ab} \tag{1.A.3}$$

where

$$R = \frac{v_a - v_b}{i_{ab}} \tag{1.A.4}$$

is called the resistance of the resistor. The unit of resistance is Ohms = Volt/Amp.

Using Eqs. (1.A.1) and (1.A.2) with Eq. (1.A.4), we obtain

$$R = \frac{v_a - v_b}{i_{ab}} = -\frac{v_b - v_a}{i_{ab}} = +\frac{v_b - v_a}{i_{ba}} = -\frac{v_a - v_b}{i_{ba}} \tag{1.A.5}$$

In 1847, Gustav Robert Kirchhoff, a professor at the University of Berlin, formulated two important laws that provide the foundation for analysis of electric circuits. These laws are referred to as *Kirchhoff's current law* (KCL) and *Kirchhoff's voltage law* (KVL) in his honor.

Kirchhoff's current law (KCL): The algebraic sum of the currents directed into a node at any instant is zero.

Kirchhoff's voltage law (KVL): The algebraic sum of the voltages around any loop in a circuit is identically zero for all time.

Kirchhoff's laws are a consequence of conservation of charge and conservation of energy [4].

The phrase *algebraic sum* in Kirchhoff's current law indicates that we must take current directions into account as we add up the currents of the elements connected to a particular node. One way to take current directions into account is to use a plus sign when the current is directed toward the

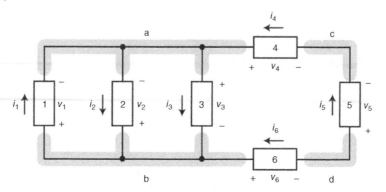

Figure 1.A.10 A circuit consisting of six elements connected together at four nodes.

node and a minus sign when the current is directed away from the node. For example, consider the circuit shown in Figure 1.A.10.

Four elements of this circuit – elements 1, 2, 3, and 4 – are connected to node a. By KCL, the algebraic sum of the element currents i_1, i_2, i_3, and i_4 must be zero. Currents i_2 and i_3 are directed away from node a, so we will use a plus sign for i_2 and i_3. In contrast, currents i_1 and i_4 are directed toward node a, so we will use a minus sign for i_1 and i_4. The KCL equation for node a of Figure 1.A.10 is

$$-i_1 + i_2 + i_3 - i_4 = 0 \qquad (1.A.6)$$

An alternate way of obtaining the algebraic sum of the currents into a node is to set the sum of all the currents directed away from the node equal to the sum of all the currents directed toward that node. Using this technique, we find that the KCL equation for node a of Figure 1.A.10 is

$$i_2 + i_3 = i_1 + i_4 \qquad (1.A.7)$$

Similarly, the KCL equation for node b of Figure 1.A.10 is

$$i_1 = i_2 + i_3 + i_6$$

KVL refers to "a *loop* in a circuit." A *loop* is a closed path through a circuit that does not encounter any intermediate node more than once. For example, starting at node a in Figure 1.A.10, we can move through element 4 to node c, then through element 5 to node d, through element 6 to node b, and finally through element 3 back to node a. We have a closed path, and we did not encounter any of the intermediate nodes – b, c, or d – more than once. Consequently, elements 3, 4, 5, and 6 comprise a loop.

Similarly, elements 1, 4, 5, and 6 comprise a loop of the circuit shown in Figure 1.A.10. Elements 1 and 3 comprise yet another loop of this circuit. The circuit has three other loops: elements 1 and 2, elements 2 and 3, and elements 2, 4, 5, and 6.

The phrase algebraic sum indicates that we must take polarity into account as we add up the voltages of elements that comprise a loop. One way to take polarity into account is to move around the loop in the clockwise direction while observing the polarities of the element voltages. We write the voltage with a plus sign when we encounter the + of the voltage polarity before the −. In contrast, we write the voltage with a minus sign when we encounter the − of the voltage polarity before the +. For example, consider the circuit shown in Figure 1.A.10. Elements 3, 4, 5, and 6 comprise a loop of the circuit. By KVL, the algebraic sum of the element voltages v_3, v_4, v_5, and v_6 must be zero. As we move around the loop in the clockwise direction, we encounter the + of v_4 before the −, the − of v_5 before the +, the − of v_6 before the +, and the − of v_3 before the +. Consequently, we use a minus sign for v_3, v_5, and v_6 and a plus sign for v_4. The KVL equation for this loop of Figure 1.A.10 is

$$v_4 - v_5 - v_6 - v_3 = 0$$

Similarly, the KVL equation for the loop consisting of elements 1, 4, 5, and 6 is

$$v_4 - v_5 - v_6 + v_1 = 0$$

The KVL equation for the loop consisting of elements 1 and 2 is

$$-v_2 + v_1 = 0$$

1.A.3 Verifying LTspice Simulation Results

Ohm's and Kirchhoff's laws provide a way to verify that we have simulated a circuit correctly. As an example, consider the circuit shown in Figure 1.A.11a, the corresponding LTspice schematic shown in Figure 1.A.11b, and simulation results shown Figure 1.A.11c. The first two lines of the simulation results indicate that the value of the node voltage at node a is 40 V and the value of the node voltage at node b is 45 V. These values are used in Figure 1.A.11d to label the nodes of the circuit with the values of their node voltages.

(Recall that Figure 1.A.3 shows how to measure the value of a node voltage using a voltmeter. The black probe of the voltmeter is connected to the ground node, and the red probe is connected to the node at which the node voltage is being measured. The voltmeter displays the value of the node voltage at the node to which the red probe is connected.)

In contrast, consider the currents in the 25 and 75 Ω resistors. The simulation results provided by LTspice specify the values of these currents but do not specify the directions of these currents. Instead, the directions of the resistor currents are determined from the simulation results using two consequences of Ohm's law:

1) The resistor current and voltage related by Ohm's law must adhere to the passive convention so the resistor current is directed from the node near the plus sign of the resistor voltage polarity toward node near the minus sign.
2) When the resistance is positive, and it almost always is, the resistor voltage and resistor current related by Ohm's law are either both positive or both negative.

Consider the 25 Ω resistor. The simulation results indicate that the resistor current is 0.2 A. This current is positive and the resistance, 25 Ω, is also positive so, using Ohm's law, the resistor voltage is also positive:

$$25(0.2) = 5 \text{ V}$$

The voltages at the nodes of the 25 Ω resistor are 40 and 45 V so the voltage across the 25 Ω resistor is $+45 - 40 = 5$ V rather than $+ 40 - 45 = -5$ V. The polarity of the voltage across the 25 Ω resistor will have the $+$ near the node at which the node voltage is 45 V and the $-$ near the node at which the node voltage is 40 V. The current in the 25 Ω resistor must be

$$\frac{5 \text{ V}}{2 \text{ Ω}} = 0.2 \text{ A}$$

directed from left-to-right to adhere to the passive convention with the resistor voltage.

Next consider the 75 Ω resistor. The simulation results indicate that the resistor current is 0.6 A. This current is positive and the resistance, 75 Ω, is also positive so, using Ohm's law, the resistor voltage is also positive:

$$75(0.6) = 45 \text{ V}$$

The voltages at the nodes of the 75 Ω resistor are 45 and 0 V so the voltage across the 75 Ω resistor is 45 V. The polarity of the voltage across the 75 Ω resistor will have the + near the top node and the − near the bottom node. The current in the 75 Ω resistor must be

$$\frac{45 \text{ V}}{75 \text{ Ω}} = 0.6 \text{ A}$$

directed downward to adhere to the passive convention with the resistor voltage.

Having verified that the node voltages and device currents obtained by simulating the circuit shown in Figure 1.A.11 satisfy both Ohm's and Kirchhoff's laws, we are now confident that the simulation results are correct.

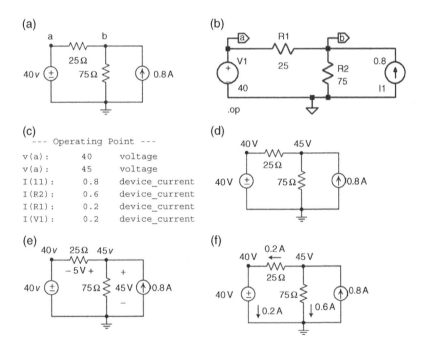

Figure 1.A.11 Interpreting LTspice simulation results. A DC circuit (a) is represented by an LTspice schematic (b). Simulating the schematic using LTspice produces node voltages and device currents tabulated in (c). Next, label the node voltages on the circuit diagram (d) and then determine the resistor voltages from the node voltages (e). The resistor currents, including the directions of the resistor currents, are determined using Ohm's law (f).

References

1 Perry, T.S. (1998). Donald Pederson: The Father of SPICE. *IEEE Spectrum* 35 (6): 22–27.

2 Mike Engelhardt (2022). Analog devices. Linear technology. LTspice, 17 August. https://en.wikipedia.org/wiki/LTspice (accessed 23 October 2022).

3 Svoboda, J.A. and Dorf, R.C. (2014). *Introduction to Electric Circuits*, 9e, 25. NJ: Wiley.

4 Svoboda, J.A. and Dorf, R.C. (2014). *Introduction to Electric Circuits*, 9e, 56. NJ: Wiley.

Chapter 2

Analysis of DC Circuits

2.1 DC Circuits

A DC circuit is a circuit in which all the voltages and currents have constant values. The circuit shown in Figure 1.4 is a DC circuit consisting of a voltage source, a current source, and two resistors. In addition to resistors, voltage sources, and current sources, a DC can include open circuits, short circuits, dependent sources, and ideal op amps. In this chapter we will analyze DC circuits containing these components.

Example 2.1

Determine the values of the voltage, v_o, across the "open circuit" and the current, i_s, in the "short circuit" shown in Figure 2.1a. The encircled numbers are node numbers.

Step 1. Formulate a circuit analysis problem.

We want to determine the values of the voltage, v_o, and the current, i_s. The voltage, v_o, is the voltage across an open circuit. An open circuit is an empty space in the circuit that is of interest. In this case the open circuit is the empty space between node 4 and node 5, and that empty space is of interest because v_o is the voltage across it.

Similarly, the current, i_s, is the current in a short circuit. A short circuit is a wire in the circuit that is of interest. In this case the short circuit is the wire between node 2 and node 3, and that wire is of interest because i_s is the current in it.

LTspice does not have built-in components called "open circuit" or "short circuit." Components called "open circuit" or "short circuit" are

---

(Continued)

LTspice® for Linear Circuits, First Edition. James A. Svoboda.
© 2023 John Wiley & Sons, Inc. Published 2023 by John Wiley & Sons, Inc.

Example 2.1   (Continued)

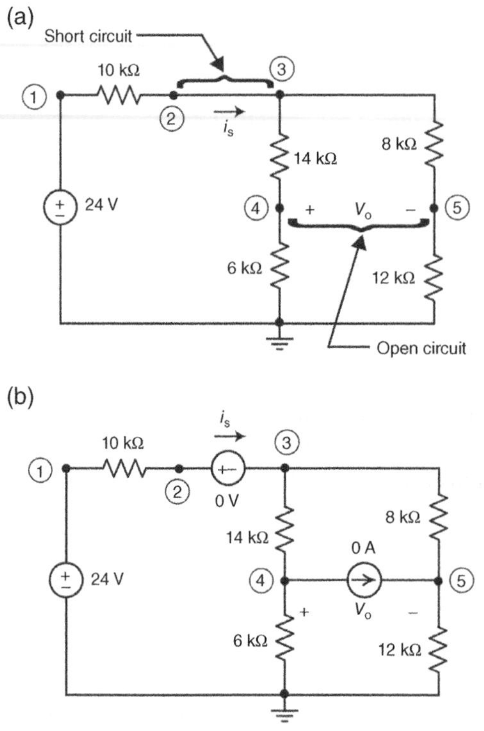

(a)

(b)

Figure 2.1   The circuit considered in Example 2.1.

unnecessary because a 0-V voltage source is equivalent to a short circuit and a 0-A current source is equivalent to an open circuit. Figure 2.1b shows the circuit from Figure 2.1a after replacing the short circuit by a 0-V voltage source and replacing the open circuit by a 0-A current source.

Step 2. Describe the circuit using an LTspice schematic.

Click on the New Schematic toolbar icon, 🔲 , to start a new schematic. Place the component symbols and adjust the values of the circuit parameters. Place a ground symbol and wire the circuit. Add node labels to obtain the schematic shown in Figure 2.2.

Step 3. Simulate the circuit using LTspice.

The circuit in Figure 2.2 is a DC circuit because the values of all the independent voltage-source voltages and the current-source currents are constants. Recall that SPICE refers to analyzing a DC circuit as "finding the DC operating point" of the circuit.

To specify the desired type of simulation, click on the Edit Tab and then select "Spice Analysis" (see Figure 1.3) to pop up the Edit Simulation Command dialog box (see Figure 1.15).

Example 2.1 (Continued)

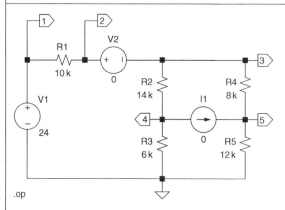

Figure 2.2 The schematic corresponding to the circuit in Figure 2.1b.

Select the "DC op pnt" tab and then click the "OK" button. LTspice will generate a "spice directive" consisting of the command ".op." Position the spice directive on the schematic as desired.

Click on the Run LTspice toolbar icon, 🏃, to perform a DC analysis of the circuit.

Step 4. Display the results of the simulation.

The results of a DC analysis display automatically as shown in Figure 2.3. The results of the DC analysis are labeled as the "Operating Point" in Figure 2.3. The operating point consists of node voltages in Volts and the circuit element currents in Amps.

Step 5. Verify that the simulation results are correct.

The node voltages and device currents shown in Figure 2.3 are correct if, and only if, they satisfy both Ohm's and Kirchhoff's laws.

We can use Ohm's law to verify that the resistor current values are consistent with the node voltage values. For example, the currents **directed from left to right** in the 10 kΩ resistor and **downward** in the 8 kΩ resistor are given by

$$\frac{V(1)-V(2)}{10\,000}=\frac{24-12}{10\,000}=0.0012\ \text{A} \quad\text{and}\quad \frac{V(3)-V(5)}{8000}=\frac{12-7.2}{8000}=0.0006\ \text{A}$$

Apparently, the current in R1 reported in the simulation results shown in Figure 2.3 is the current directed from right to left in the 10 kΩ resistor, and the current reported for R4 is the current directed upward in the 8 kΩ resistor. (The orientation of device currents reported by

(Continued)

Example 2.1 (Continued)

```
*⌐  * C:\Program Files (x86)\LTC\LTspiceIV\Example2.1.asc        ⊠
         --- Operating Point ---

V(1):           24              voltage
V(2):           12              voltage
V(4):           3.6             voltage
V(3):           12              voltage
V(5):           7.2             voltage
I(I1):          0               device_current
I(R5):          -0.0006         device_current
I(R4):          -0.0006         device_current
I(R3):          -0.0006         device_current
I(R2):          -0.0006         device_current
I(R1):          -0.0012         device_current
I(V2):          0.0012          device_current
I(V1):          -0.0012         device_current
```

Figure 2.3   Simulation results.

LTspice is discussed in Example 1.1.) We can use the simulation results to label the node voltages and resistor currents on the circuit diagram as shown in Figure 2.4. We readily verify that the currents given in the simulation results satisfy Kirchhoff's current law.

Step 6. Report the result.

The value of v_o, the voltage across open circuit in Figure 2.1, is equal to the voltage across the 0-A current source in Figure 2.4, which is

$$v_o = V(4) - V(5) = 3.6 - 7.2 = -3.6 \, V$$

The value of i_s, the current in the short circuit in Figure 2.1, is equal to the current in the 0-V current source in Figure 2.4, that is

$$i_s = 1.2 \, mA$$

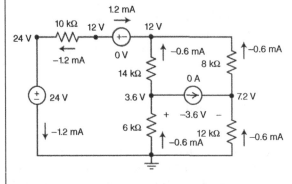

Figure 2.4   Node voltages and resistor currents.

2.2 Dependent Sources

Dependent sources consist of two parts [1]. One part, called the controlling part, is either a short circuit or an open circuit. The other part, called the controlled part, is either a voltage source or a current source. The voltage of the controlled voltage source or current of the controlled current source is proportional to the current in the short circuit or the voltage across the open circuit. In SPICE the four types of dependent sources are represented by E, F, G, and H components. Table 2.1 describes dependent sources and their representations in LTspice.

Table 2.1 Dependent sources and corresponding LTspice notation.

Circuit diagram name	Circuit diagram symbol	LTspice component name	LTspice component symbol
Voltage Controlled Voltage Source		E	
Voltage Controlled Current Source		G	
Current Controlled Voltage Source		H	
Current Controlled Current Source		F	

Example 2.2

Determine the values of the current i_2 and voltage v_3 in the circuit shown in Figure 2.5a.

Step 1. Formulate a circuit analysis problem.

The circuit in Figure 2.5a contains two dependent sources: a voltage-controlled current source and a current-controlled voltage source. Table 2.1 summarizes the symbols and notation used to represent dependent sources, both in circuit diagrams and in LTspice schematics. Notice from Table 2.1 that, in LTspice, a controlling voltage of a dependent source is required to be the voltage across an open circuit and a controlling current is required to be the current in a short circuit. In contrast, in Figure 2.5a the controlling current of the dependent voltage source is the current in the $10\,\Omega$ resistor and the controlling voltage of the dependent current source is the voltage across the $20\,\Omega$ resistor. In anticipation of using LTspice, we redraw the circuit of Figure 2.5a as shown in Figure 2.5b. The circuit in Figure 2.5b both is equivalent to the circuit in Figure 2.5a and satisfies the requirements of Table 2.1.

The current i_2 in a short circuit in Figure 2.5b is the same current as the current i_2 in the $10\,\Omega$ resistor in Figure 2.5a. Similarly, the voltage v_3 across an open circuit in Figure 2.5b is the same voltage as the voltage v_3 across the $20\,\Omega$ resistor in Figure 2.5a. This is important because

(a)

(b)

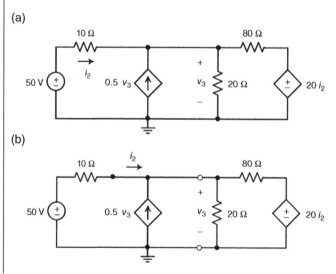

Figure 2.5 The circuit considered in Example 2.2.

Example 2.2 (Continued)

Figure 2.5b conforms to the requirements of Table 2.1 and Figure 2.5a does not conform to the requirements of Table 2.1. The circuit in Figure 2.5b is equivalent to the circuit in Figure 2.5a.

Step 2. Describe the circuit using an LTspice schematic.

Click on the New Schematic toolbar icon, , to start a new schematic. Place symbols representing the 50-V voltage source and the three resistors. Adjust the values of the circuit parameters. Place a ground symbol and wire the circuit. Add node labels to obtain the schematic shown in Figure 2.6.

Next, consider the voltage-controlled current source in the circuit in Figure 2.5b. Table 2.1 indicates that the LTspice component "G" represents a voltage-controlled current source. Compare Table 2.1 and Figure 2.5b to see that v_c is the voltage across the 20 Ω resistor and the value of the gain of the voltage-controlled voltage source is $k = 0.5$ A/V.

Click on the Component icon, , on the toolbar to pop up the Select Component Symbol dialog box. Type "G" in the search box. LTspice will find the symbol for a voltage-controlled current source. Click the "OK" button on the Select Component Symbol dialog box. Position the voltage-controlled current source symbol and wire it into the schematic as shown in Figure 2.7.

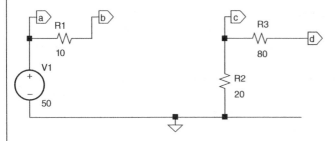

Figure 2.6 Drawing the schematic for the circuit of Figure 2.5b – Stage 1.

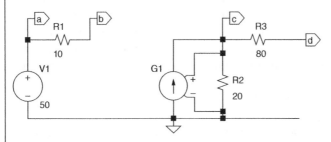

Figure 2.7 Drawing the schematic for the circuit of Figure 2.5b – Stage 2.

(Continued)

Example 2.2   (Continued)

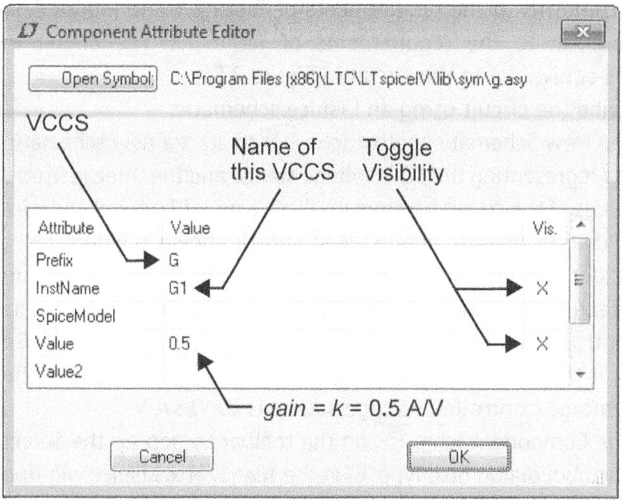

Figure 2.8   Specifying the gain of the voltage-controlled current source.

This may require rotating and/or reflecting an LTspice symbol. Select a symbol and type <control><R> to rotate the symbol clockwise 90°. Similarly, select a symbol and type <control><E> to mirror the symbol.

Next, right-click on the LTspice symbol for component G1 to pop up the "Component Attribute Editor" dialog box. Enter the value of the gain, k, in A/V and mark the name of the dependent source and its gain to be visible as shown in Figure 2.8.

Next, consider the current-controlled voltage source in the circuit in Figure 2.5b. Table 2.1 indicates that the LTspice representation of a current-controlled voltage source consists of two parts: a short circuit implemented as a 0-V voltage source and an "H" component. Compare Table 2.1 and Figure 2.5b to see that i_2 is the current in the 10 Ω resistor and the value of the gain k of the current-controlled voltage source is 20.

Add a voltage source to the schematic and set its voltage to zero as shown in Figure 2.9.

Click on the Component icon, ⊃, on the toolbar to pop up the Select Component Symbol dialog box. Type "H" in the search box. LTspice will find the symbol for a voltage-controlled current source. Click the "OK" button on the Select Component Symbol dialog box.

Example 2.2 (Continued)

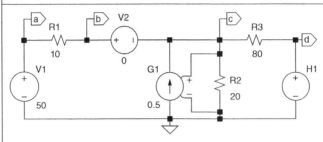

Figure 2.9 Drawing the schematic for the circuit of Figure 2.5b – Step 3.

Position the short circuit (labeled V2 in Figure 2.9) and the controlled voltage source symbol (labeled H1 in Figure 2.9), and wire them into the schematic as shown in Figure 2.9.

Right-click on the LTspice symbol for component H1 to pop up the "Component Attribute Editor" dialog box shown in Figure 2.10. Enter the name of the short circuit, the value of the gain, k, in V/A as shown in Figure 2.10. Mark the name of the dependent source, the name of the

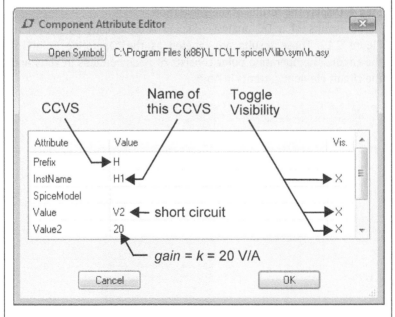

Figure 2.10 Specifying the gain of the current-controlled voltage source.

(Continued)

Example 2.2 (Continued)

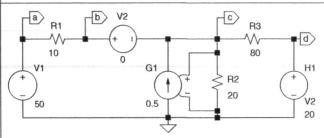

Figure 2.11 The completed schematic corresponding to the circuit in Figure 2.5b.

short circuit, and the gain to be visible as shown in Figure 2.10. Figure 2.11 shows the completed schematic.

Step 3. Simulate the circuit using LTspice.

Click on the Edit Tab, and then select "Spice Analysis" to pop up the Edit Simulation Command dialog box. Select the "DC op pnt" tab, and then click the "OK" button. LTspice will generate a "spice directive" consisting of the command ".op." Position the spice directive on the schematic as desired.

Click on the Run LTspice toolbar icon, 🏃, to simulate the circuit.

Step 4. Display the results of the simulation.

The results of a DC analysis display automatically as shown in Figure 2.12. These results are labeled as the "Operating Point" of the circuit. The operating point consists of node voltages in Volts and the circuit element currents in Amps.

```
LT * C:\Program Files (x86)\LTC\LTspiceIV\Exa2_2.asc          [X]

          --- Operating Point ---

V(a):          50             voltage
V(b):          -20            voltage
V(c):          -20            voltage
V(d):          140            voltage
I(H1):         -2             device_current
I(R3):         2              device_current
I(R2):         -1             device_current
I(R1):         -7             device_current
I(G1):         -10            device_current
I(V2):         7              device_current
I(V1):         -7             device_current
```

Figure 2.12 Simulation results.

Example 2.2 (Continued)

Step 5. Verify that the simulation results are correct.

We can use the simulation results to label the node voltages and resistor currents on the circuit diagram as shown in Figure 2.13. The simulation results are correct if, and only if, these node voltages and device currents satisfy Ohm's and Kirchhoff's laws and the equations describing the two dependent sources.

Let us check that the current in the dependent current source is indeed given by $0.5v_3$, as required by Figure 2.5. Noticing that v_3 is equal to the node voltage at node b, we can calculate the current in the dependent current source as

$$0.5V(b) = 0.5(-20) = -10\,A$$

where V(b) denotes the node voltage at node b.

Next let us check that the voltage of the dependent voltage source is indeed given by $20\,i_2$.

Noticing that voltage across the dependent voltage source is equal to the node voltage at node c and that i_2 is the current in 0-V voltage source V2, we can calculate the voltage across the dependent voltage source as

$$140\,V = V(c) = 20(I(V2)) = 20(7)$$

Both these equations are correct. Also, we readily verify that the currents given in the simulation results satisfy Kirchhoff's current law at nodes a, b, and c. We conclude that the simulation results are correct.

Step 6. Report the result.

The values of the current i_2 and of the voltage v_3 in the circuit shown in Figure 2.5a are

$$i_2 = 7\,A \quad \text{and} \quad v_3 = -20\,V$$

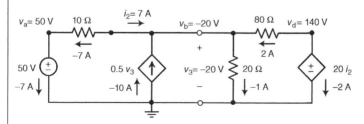

Figure 2.13 Node voltages and device currents.

Example 2.3

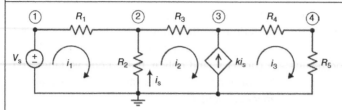

Figure 2.14 The circuit considered in Example 2.3.

Consider the circuit shown in Figure 2.14. A set of mesh currents has been labeled and the nodes of this circuit have been numbered.

Suppose that this circuit has been represented by the following mesh equations [2].

$$(R_1 + R_2)i_1 - R_2i_2 = V_s$$
$$-R_2i_1 + (R_2 + R_3)i_2 + (R_4 + R_5)i_3 = 0$$
$$-ki_1 + (k+1)i_2 - i_3 = 0$$

The objective of this example is to use LTspice to determine if these mesh equations are correct.

Step 1. Formulate a circuit analysis problem.

We modify the circuit from Figure 2.14 in three ways in anticipation of using LTspice to determine the values of the mesh currents, i_1, i_2, and i_3.

1) In Figure 2.14 the resistances, voltage-source voltage, and gain of the dependent source have symbolic values. Numerical values of these circuit parameters are required for analysis using LTspice. We chose, somewhat randomly, the values

$$R_1 = 200\,\Omega, R_2 = 500\,\Omega, R_3 = 250\,\Omega, R_4 = 50\,\Omega,$$
$$R_5 = 200\,\Omega, V_s = 40\text{ V}, k = 4\text{ A/A}$$

With these values the mesh equations become

$$700i_1 - 500i_2 = 40$$
$$-500i_1 + 750i_2 + 250i_3 = 0$$
$$-4i_1 + 5i_2 - i_3 = 0$$

and the mesh currents become

$$i_1 = 0.12\,\text{A}, i_2 = 0.088\,\text{A}, \text{and } i_3 = -0.04\,\text{A}$$

Example 2.3 (Continued)

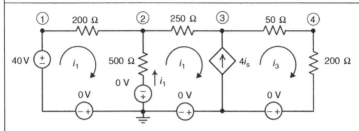

Figure 2.15 The circuit from Figure 2.14 after adding short circuits (0-V voltage sources) to measure the mesh currents.

2) A wire carrying the controlling current of the dependent source, i_s, is represented as a short circuit implemented as a 0-V voltage source.
3) Wires carrying the mesh currents are each represented as a short circuit implemented as a 0-V voltage source.

The modified circuit is shown in Figure 2.15.

Step 2. Describe the circuit using an LTspice schematic.

Click on the New Schematic toolbar icon, 🔲, to start a new schematic. Place the component symbols and adjust the values of the circuit parameters.

Table 2.1 indicates that the current-controlled current source consists of two parts: a 0-V voltage source and an "F" component. Right-click on the symbol of the "F" component to pop up the dialog box shown in Figure 2.16 and specify the attributes of that "F" component. Place a

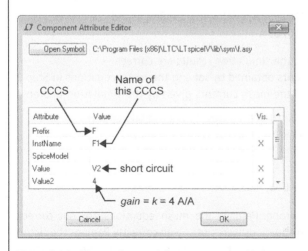

Figure 2.16 The attributes of the current-controlled current source.

(Continued)

Example 2.3   (Continued)

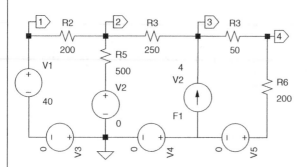

Figure 2.17 An LTspice schematic corresponding to the circuit in Figure 2.15.

ground symbol and wire the circuit. Add node labels to obtain the schematic shown in Figure 2.17.

Step 3. Simulate the circuit using LTspice.

Click on the Edit Tab, and then select "Spice Analysis" to pop up the Edit Simulation Command dialog box. Select the "DC op pnt" tab, and then click the "OK" button. LTspice will generate a "spice directive" consisting of the command ".op." Position the spice directive on the schematic as desired.

Click on the Run LTspice toolbar icon, ✗, to simulate the circuit.

Step 4. Display the results of the simulation.

The simulation results indicate that the mesh currents are

$$I(V3) = 0.0461538\,A, I(V4) = -0.0153846\,A, \text{ and}$$
$$I(V5) = 0.230769\,A$$

Step 5. Verify that the simulation results are correct.

The mesh currents obtained by solving the mesh equations in Step 1 are different than the mesh currents given by the simulation in Step 4 results:

$$0.12 = i_1 \neq I(V3) = -0.046\,A, 0.088 = i_2 \neq I(V4) = -0.015\,A,$$

and

$$-0.04 = i_3 \neq I(V5) = 0.023\,A$$

Something is wrong. Perhaps the mesh equations are not correct. Perhaps the mesh equations were not solved correctly. Perhaps the circuit was not simulated correctly. After reviewing the preceding steps 1 and 2,

Example 2.3 (Continued)

we notice that the polarity of voltage source V2 in the LTspice schematic in Figure 2.17 is different than the polarity of the corresponding voltage source in the circuit diagram in Figure 2.15. That difference is enough to cause the simulation to produce incorrect values of the mesh currents.

Figure 2.18 shows the corrected schematic. Rerunning the simulation produces the expected values of the mesh currents:

$$I(V3) = i_1 = 0.12\,\text{A}, \ \ I(V4) = i_2 = 0.088\,\text{A}, \ \text{and} \ \ I(V5) = i_3 = -0.04\,\text{A}$$

We conclude that both the corrected simulation and the mesh equations are correct.

Step 6. Report the answer to the circuit analysis problem.
The equations

$$\left(R_1 + R_2\right)i_1 - R_2 i_2 = V_s$$
$$-R_2 i_1 + \left(R_2 + R_3\right)i_2 + \left(R_4 + R_5\right)i_3 = 0$$
$$-ki_1 + \left(k+1\right)i_2 - i_3 = 0$$

are a correct set of mesh equations representing the circuit shown in Figure 2.14.

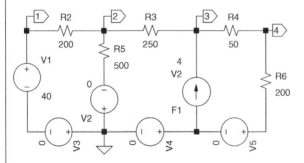

Figure 2.18 The LTspice schematic after correcting the polarity of voltage source V2.

2.3 Equivalent Circuits

It is often useful to replace part of a circuit with an "equivalent circuit." Consider the circuit shown in Figure 2.19a. This circuit is divided into two parts by the pair of terminals labeled "A" and "B." In Figure 2.19b the part

(a)

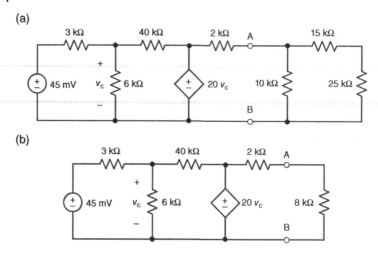

(b)

Figure 2.19 A circuit (a) and an equivalent circuit (b).

of the circuit to the right of the terminals has been replaced by an "equivalent circuit" [3]. We say that the 8 kΩ resistor connected between nodes A and B in Figure 2.19b is equivalent to the combination of the 10, 15, and 25 kΩ resistors connected between nodes A and B in Figure 2.19a. By this we mean that replacing the combination of the 10, 15, and 25 kΩ resistors in Figure 2.19a by an 8 kΩ resistor does not change the current or voltage of any component in the rest of the circuit.

LTspice provides an easy way to confirm that 8 kΩ resistor in Figure 2.19b is equivalent to the combination of the 10, 15, and 25 kΩ resistors in Figure 2.19a. We simulate circuits and check that the current and voltage of each component to the left of terminals labeled "A" and "B" is the same in both the simulations.

Example 2.4

Confirm that 8 kΩ resistor in Figure 2.19b is equivalent to the combination of the 10, 15, and 25 kΩ resistors in Figure 2.19a.

Step 1. Formulate a circuit analysis problem.
   Simulate the circuits shown in Figure 2.19 and check that the current and voltage of each component to the left of terminals labeled "A" and "B" is the same in both the simulations.

Step 2. Describe the circuit using an LTspice schematic.
   Let us prepare two separate schematics. First, prepare a schematic corresponding to the circuit diagram in Figure 2.19a. Click

Example 2.4   (Continued)

(a)

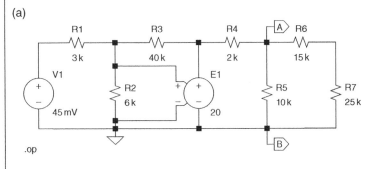

.op

(b)

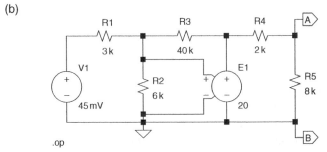

.op

Figure 2.20   The schematics corresponding to the circuit in Figure 2.19a,b.

on the New Schematic toolbar icon, 🔲, to start a new schematic. Place the component symbols and adjust the values of the circuit parameters. Place a ground symbol and wire the circuit to obtain the schematic shown in Figure 2.20a. Save this schematic as Exa2_4a.asc.

Step 3. Simulate the circuit using LTspice.

Open the schematic in file Exa2_4a.asc.

Select the "DC op pnt" tab, and then click the "OK" button. LTspice will generate a "spice directive" consisting of the command ".op." Position the spice directive on the schematic as desired.

Click on the Run LTspice toolbar icon, 🏃, to simulate the circuit. The simulation results are shown in column (a) of Figure 2.21.

Open the schematic in file Exa2_4b.asc. Request the calculation of the operating point of a DC circuit and run that simulation. The simulation results are shown in column (a) of Figure 2.21.

(*Continued*)

Example 2.4   (Continued)

```
     --- Operating Point ---                 --- Operating Point ---
V(n001):   0.045 voltage            V(n001):   0.045      voltage
V(n002):   0.6   voltage            V(n002):   0.6        voltage
V(n003):   12    voltage            V(n003):   12         voltage
V(n004):   9.6   voltage            V(n004):   9.6        voltage
V(n005):   6     voltage
I(R7):   0.00024   device_current
I(R6):   0.00024   device_current
I(R5):   0.00096   device_current   I(R5):   0.0012     device_current
I(R4):   0.0012    device_current   I(R4):   0.0012     device_current
I(R3):  -0.000285  device_current   I(R3):  -0.000285   device_current
I(R2):   0.0001    device_current   I(R2):   0.0001     device_current
I(R1):  -0.000185  device_current   I(R1):  -0.000185   device_current
I(E1):  -0.001485  device_current   I(E1):  -0.001485   device_current
I(V1):   0.000185  device_current   I(V1):   0.000185   device_current
```

(a)   (b)

Figure 2.21   Simulation results.

Step 4. Display the results of the simulation.

Step 5. Verify that the simulation results are correct.

Consider the part circuit to the left of the terminals in Figure 2.20. Replacing the combination of the 10, 15, and 25 kΩ resistors by the 8 kΩ "equivalent resistor" did not change any node voltage or device current in that part of the circuit.

Step 6. Report the result.

The 8 kΩ resistor in Figure 2.20b is equivalent to the combination of the 10, 15, and 25 kΩ resistors in Figure 2.20a.

2.4   Thevenin Equivalent Circuits

It is remarkably easy to use LTspice to find a Thevenin equivalent circuit [4]. Given a circuit:

a) Insert a pair of terminals to mark the part of the circuit that is to be replaced by its Thevenin equivalent circuit.

b) Remove the rest of the circuit. Simulate the part of the circuit that is to be replaced by its Thevenin equivalent circuit twice: once to determine the value of the open circuit voltage, v_{oc}, and once to determine the value of the short circuit current, i_{sc}.

c) Determine the value of the Thevenin resistance, R_t, using, $v_{oc} = R_t i_{sc}$.

d) The Thevenin equivalent circuit consists of the series connection of a voltage source having voltage v_{oc} and a resistor having resistance R_t.

Example 2.5

Consider the circuit shown in Figure 2.22. Can the value of the resistance R be selected so that $v_R = 10\,\text{V}$? If yes, what is the required value of R?

Step 1. Formulate a circuit analysis problem.

Consider the part of the circuit to the left of the terminals in Figure 2.22. Figure 2.23 shows how the open circuit voltage, v_{oc}, and the short circuit current, i_{sc}, of this part of the circuit are calculated. We determine the values of v_{oc} and i_{sc} by simulating these circuits using LTspice. We can then replace the part of the circuit in Figure 2.22 to the left of the terminals by its Thevenin equivalent circuit, as shown in Figure 2.24, and answer the questions regarding the value of the resistance R.

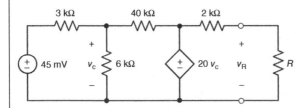

Figure 2.22 The circuit considered in Example 2.5.

(a)

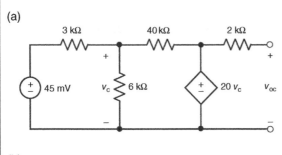

(b)

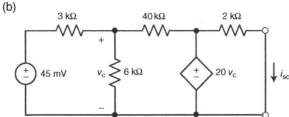

Figure 2.23 The (a) open circuit voltage, v_{oc}, and (b) short circuit current, i_{sc}.

(Continued)

Example 2.5   (Continued)

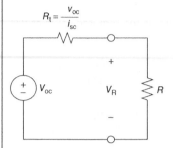

Figure 2.24   Replacing the part of the circuit by its Thevenin equivalent circuit.

Step 2. Describe the circuit using an LTspice schematic.

Figure 2.25a shows the LTspice schematic corresponding to Figure 2.23a. The open circuit in Figure 2.23a is represented as a 0-A current source in Figure 2.25a. The voltage across that 0-A current source is the open circuit voltage, v_{oc}.

Figure 2.25b shows the LTspice schematic corresponding to Figure 2.23b. The short circuit in Figure 2.23b is represented as a 0-V voltage source in Figure 2.25b. The current in that 0-V voltage source is the short circuit current, i_{sc}.

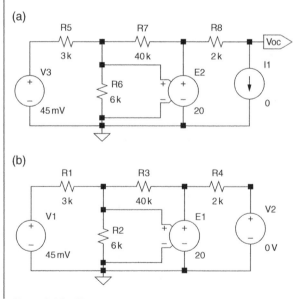

Figure 2.25   The schematics used to determine (a) open circuit voltage, v_{oc}, and (b) short circuit current, i_{sc}.

Example 2.5 (Continued)

Step 3. Simulate each circuit using LTspice.

Simulate the schematic in Figure 2.25a to determine the operating point of this DC circuit. The simulation results are shown in column (a) of Figure 2.26.

Simulate the schematic in Figure 2.25b to determine the operating point of this DC circuit. The simulation results are shown in column (b) of Figure 2.26.

(a)   (b)

```
--- Operating Point ---                --- Operating Point ---
V(n001):  0.045      voltage      V(n001):  0.045  voltage
V(n002):  0.6        voltage      V(n002):  0.6    voltage
V(n003):  12         voltage      V(n003):  12     voltage
V(voc):   12         voltage      V(n004):  0      voltage

I(I1):    0          device_current   I(R4):  0.006      device_current
I(R8):    0          device_current   I(R3): -0.000285   device_current
I(R7):   -0.000285   device_current   I(R2):  0.0001     device_current
I(R6):    0.0001     device_current   I(R1): -0.000185   device_current
I(R5):   -0.000185   device_current   I(E1): -0.006285   device_current
I(E2):   -0.000285   device_current   I(V2):  0.006      device_current
I(V3):    0.000185   device_current   I(V1):  0.000185   device_current
```

Figure 2.26 Simulation results.

Step 4. Display the results of the simulation.

The simulation shown in Figure 2.26 results indicate that

$$v_{oc} = V(voc) = 12\,V \text{ and } i_{sc} = I(V2) = 0.006 = 6\,mA$$

The value of the Thevenin resistance, R_t, is determined from the values of v_{oc} and i_{sc} to be

$$R_t = \frac{v_{oc}}{i_{sc}} = \frac{12}{0.006} = 2000\,\Omega$$

Referring to Figure 2.24, we calculate the value of the resistance R required to cause $v_R = 10$ V:

$$10 = v_R = \frac{R}{R + R_t}v_{oc} = \frac{R}{R + 2000}(12) \quad \Rightarrow \quad R = \frac{2000(10)}{12 - 10} = 10\,k\Omega$$

Step 5. Verify that the simulation results are correct.

An LTspice simulation of the circuit in Figure 2.22 with $R = 10\,k\Omega$ verifies that $v_R = 10$ V when $R = 10\,k\Omega$.

Step 6. Report the result: Select $R = 10\,k\Omega$ in Figure 2.22 to cause $v_R = 10$ V.

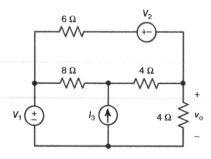

Figure 2.27 The circuit considered in Example 2.6.

2.5 Superposition

The inputs to the circuit shown in Figure 2.27 are the voltage-source voltages and the current-source current. The output is the voltage v_o across one of the 4Ω resistors. The principle of superposition [5] requires that there exist constants a_1, a_2, and a_3 such that

$$v_o = a_1 V_1 + a_2 V_2 + a_3 I_3 \qquad (2.1)$$

We can easily determine the values of a_1, a_2, and a_3 by choosing convenient values of the inputs V_1, V_2, and I_3 and measuring the corresponding value of the output v_o:

$$
\begin{aligned}
v_o = a_1 \quad &\text{when} \quad V_1 = 1\text{ V, } V_2 = 0,\text{ and } I_3 = 0 \\
v_o = a_2 \quad &\text{when} \quad V_1 = 0,\text{ } V_2 = 1\text{ V, and } I_3 = 0 \\
v_o = a_3 \quad &\text{when} \quad V_1 = 0,\text{ } V_2 = 0,\text{ and } I_3 = 1\text{ A}
\end{aligned}
\qquad (2.2)
$$

Example 2.6

Use superposition to analyze the circuit in Figure 2.27.

Step 1. Formulate a circuit analysis problem.
   Determine the values of the constants a_1, a_2, and a_3 required to represent the circuit in Figure 2.27 by Eq. (2.1).

Step 2. Describe the circuit using an LTspice schematic.
   Place the component symbols and enter the values of the resistances. Place a ground symbol and wire the circuit. Label the output node to obtain the schematic shown in Figure 2.28.
   Enter values for the voltage-source voltages and the current-source current. Begin with

$$V_1 = 1\text{ V, } V_2 = 0,\text{ and } I_3 = 0$$

Step 3. Simulate the circuit using LTspice.
   Click on the New Schematic toolbar icon, ⌨, to start a new schematic. Click on the Edit Tab, and then select "Spice Analysis" to pop up the Edit Simulation Command dialog box. Select the "DC op pnt" tab, and then click the "OK" button. LTspice will generate a "spice directive" consisting of the command ".op." Position the spice directive on the

Example 2.6   (Continued)

Figure 2.28 An LTspice schematic corresponding to the circuit in Figure 2.27.

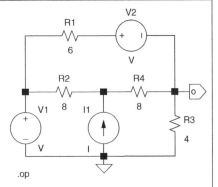

schematic as desired. Click on the Run LTspice toolbar icon, 🏃, to simulate the circuit.

Step 4. Display the results of the simulation.

The three simulations results are shown in Figure 2.29. The top row of Figure 2.29 displays the simulation results for the case in which $V_1 = 1\,V, V_2 = 0$, and $I_3 = 0$. In this case V(o) = a_1 so $a_1 = V(o) = \dfrac{1}{2}$.

Edit values for the voltage-source voltages and the current-source current so that $V_1 = 0\,V, V_2 = 1$, and $I_3 = 0$. Click on the Run LTspice toolbar icon, 🏃, to simulate the circuit with these values.

The middle row of Figure 2.29 displays the simulation results for the case in which $V_1 = 0, V_2 = 1\,V$, and $I_3 = 0$. In this case V(o) = a_2 so $a_2 = V(o) = -\dfrac{1}{3}$.

Edit values for the voltage-source voltages and the current-source current so that $V_1 = 1\,V, V_2 = 0$, and $I_3 = 0$. Click on the Run LTspice toolbar icon, 🏃, to simulate the circuit with these values.

The bottom row of Figure 2.29 displays the simulation results for the case in which $V_1 = 0, V_2 = 0\,V$, and $I_3 = 1\,A$. In this case V(o) = a_3 so $a_3 = V(o) = \dfrac{4}{3}$.

The output voltage of the circuit shown in Figure 2.28 is given by

$$v_0 = \frac{1}{2}V_1 - \frac{1}{3}V_2 + \frac{4}{3}I_3 \tag{2.3}$$

(Continued)

Example 2.6   (Continued)

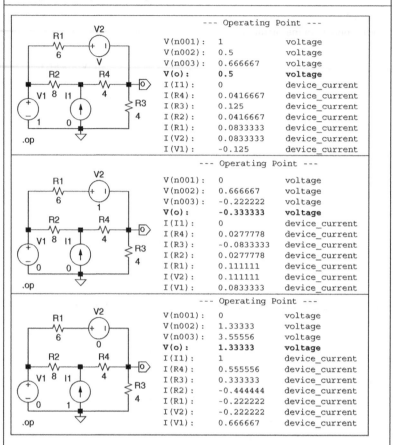

Figure 2.29   Simulation results.

Step 5. Verify that the simulation results are correct.

Let us choose, somewhat randomly, $V_1 = 5\,\text{V}, V_2 = -6\,\text{V},$ and $I_3 = 3\,\text{A}$. Using Eq. (3.3) we obtain

$$v_o = \frac{1}{2}(5) - \frac{1}{3}(-6) + \frac{4}{3}(3) = 8.5\,\text{V} \tag{2.4}$$

Example 2.6 **(Continued)**

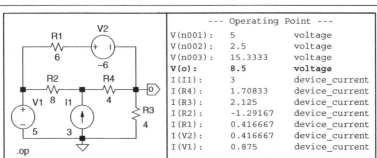

```
                              --- Operating  Point ---
V(n001):      5                 voltage
V(n002):      2.5               voltage
V(n003):      15.3333           voltage
V(o):         8.5               voltage
I(I1):        3                 device_current
I(R4):        1.70833           device_current
I(R3):        2.125             device_current
I(R2):        -1.29167          device_current
I(R1):        0.416667          device_current
I(V2):        0.416667          device_current
I(V1):        0.875             device_current
```

Figure 2.30 Simulation results with V_1 = 5 V, V_2 = –6 V, and I_3 = 3 A.

Figure 2.30 displays the results of simulating the circuit with V_1 = 5 V, V_2 = –6 V, and I_3 = 3 A. These simulation results indicate that V(o) = 8.5 V as predicted by Eq. (3.4). The simulation results seem to be correct.

Step 6. Report the result.

The output of the circuit shown in Figure 2.27 is related to the inputs of the circuit by the equation

$$v_o = \frac{1}{2}V_1 - \frac{1}{3}V_2 + \frac{4}{3}I_3 \qquad (2.5)$$

2.6   Conservation of Energy

The inputs to a circuit consist of the voltages of the voltage sources and the currents of the current sources. A DC circuit is a circuit in which all the inputs have constant values. Consequently, all the voltages and currents of a DC circuit have constant values, conventionally represented as numbers having units of volts or amps as appropriate. For historical reasons, analysis of a DC circuit using SPICE is called finding the "DC operating point" of that circuit.

Example 2.7

Consider the circuit shown in Figure 2.31. We want to determine the power supplied by each of the sources. The bottom node of the circuit is specified to be the ground node. The other three nodes are labeled as node "a," node "b," and node "c." The inputs to this circuit are the voltage of the voltage source and the current of the current source. We want to determine the power supplied by each of the sources.

The circuit diagram in Figure 2.31 provides the resistance of each resistor and the labels of the nodes to which that resistor is connected. For example, the 48 Ω resistor is connected to nodes "b" and "c." Similarly, the 12 V voltage source is connected to node "a" and "ground" with the "+" node of the voltage source connected to node "a" and the "−" node of the voltage source connected to "ground."

Step 1. Formulate a circuit analysis problem.

Determine the power supplied by each of the sources in the circuit shown in Figure 2.31. To do so, we first determine the values of the node voltages.

Step 2. Describe the circuit using an LTspice schematic.

Click on the New Schematic toolbar icon, ⟨⟩, to start a new schematic. Place the component symbols and adjust the values of the circuit parameters. Place a ground symbol and wire the circuit. Add node labels to obtain the schematic shown in Figure 2.32.

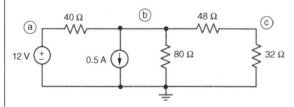

Figure 2.31   A DC circuit.

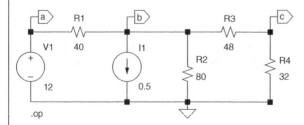

Figure 2.32   The schematic corresponding to the circuit in Figure 2.31.

Example 2.7 (Continued)

LTspice names the resistors as they are added to the schematic. For example, R_1 is the name of the 40 Ω resistor connected between node "a" and node "b," and R_2 is name of the 80 Ω resistor connected between node "b" and the ground node.

Similarly, LTspice names the voltage and current sources as they are added to the schematic. For example, V_1 is name of the 12 V voltage source connected between node "a" and the ground node.

Step 3. Simulate the circuit using LTspice.

The circuit in Figure 2.32 is a DC circuit because the values of the inputs are constants. Consequently, all the voltages and currents of this circuit have constant values. Recall that SPICE refers to analyzing a DC circuit as "finding the DC operating point" of the circuit.

To specify the desired type of simulation, click on the Edit Tab and then select "Spice Analysis" to pop up the Edit Simulation Command dialog box. Select the "DC op pnt" tab, and then click the "OK" button. LTspice will generate a "spice directive" consisting of the command ".op." Position the spice directive on the schematic as shown in Figure 2.32.

Click on the Run LTspice toolbar icon, 🏃, to simulate the circuit.

Step 4. Display the results of the simulation.

The results of a DC analysis, labeled as the "Operating Point," display automatically as shown next:

Let us make some observations about the simulation results:

1) The three nodes along the top of the circuit are labeled as node "a," node "b," and node "c" in Figure 2.31. Node labels are used in the LTspice schematic in Figure 2.32 to label the nodes in the same way. Consequently, V(a), V(b), and V(c) in the simulation results are indeed the values of the node voltages at nodes "a," "b," and "c" in the circuit shown in Figure 2.31.

2) The circuit in Figure 2.31 contains four resistors labeled only by their resistances: 40, 80, 48, and 32 Ω. Those four resistors were numbered as they were added to the schematic, and the resistors were named R1, R2, R3, and R4 as shown on the schematic in Figure 2.32.

3) Similarly, the voltage source and current source in Figure 2.31 are labeled only by the value of the voltage-source voltage and the value of the current-source current. Voltage sources are numbered as they are added to the schematic, as are current sources. This numbering is used to name the voltage source as V1 and the current source as I1 as shown in Figure 2.32.

(Continued)

Example 2.7 (Continued)

4) The labels R1, R2, R3, and R4 of the resistors and V1 and I1 of the voltage and current source are used to identify the devices' currents in the simulation results in Figure 2.33.

a) The simulation results indicate that the current in resistor R_1, the 40 Ω resistor, is I(R1) = 0.4 A, but the simulation results do not indicate whether I(R1) = 0.4 A is the current directed right-to-left or left-to-right in resistor R_1. We can use Ohm's law to determine the direction of I(R1). Resistor R_1 has a positive resistance value, as is usually the case. The following guidelines, based on Ohm's law, are used to determine the direction of resistor currents.

A resistor current having a positive value is directed from the node of that resistor having the larger node voltage toward the node of that resistor having the smaller node voltage.

A resistor current having a negative value is directed from the node of that resistor having the smaller node voltage toward the node of that resistor having the larger node voltage.

The simulation results indicate that current I(R1) is positive. Therefore, Ohm's law requires that I(R1) is directed from the node of R1 with the larger node voltage, node "a," toward the node of R1 having the smaller node voltage, node "b." Similar observations are used to determine the directions of the currents I(R2), I(R3), and I(R4).

b) The voltage-source current, I(V1) = −0.4 A, must be directed so that I(R1) and I(V1) satisfy Kirchhoff's current law at node "a."

5) Figure 2.34 shows the circuit from Figure 2.31 labeled with the simulation results from Figure 2.33

```
                --- Operating Point ---

V(a):                 12                  voltage
V(b):                 -4                  voltage
V(c):                 -2.4                voltage
I(I1):                0.5                 device_current
I(R4):                -0.05               device_current
I(R3):                -0.05               device_current
I(R2):                -0.05               device_current
I(R1):                0.4                 device_current
I(V1):                -0.4                device_current
```

Figure 2.33 Simulation results: node voltages and device currents.

Example 2.7 (Continued)

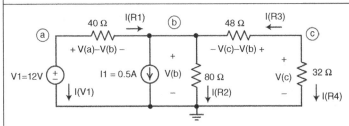

Figure 2.34 The circuit diagram from Figure 2.31, labeled with the simulation results from Figure 2.33.

Step 5. Verify that the simulation results are correct.

The terms "passive convention" and "active convention" describe the relationship of the polarity of the device voltage and the direction of the device current as shown in Figure 2.35. The device current and voltage are said to adhere to the **passive convention** when the current is directed from the node near the plus sign of the voltage polarity toward the node near the minus sign of the voltage polarity. Conversely, the element current and voltage are said to adhere to the **active convention** when the current directed from the node near the minus sign of the voltage polarity toward the node near the plus sign of the voltage polarity.

When the device voltage and the device current adhere to the passive convention, the product of that device voltage and device current is "the power supplied *to* that device *by* the rest of the circuit." On the other hand, when the device voltage and the device current adhere to the active convention, the product of that device voltage and device current is "the power supplied *by* that device *to* the rest of the circuit." Consider, for example, the voltage source in Figure 2.34. The voltage-source

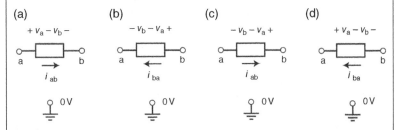

Figure 2.35 The device current and device voltage adhere to the passive convention in (a) and (b) and to the active convention in (c) and (d).

(Continued)

Example 2.7 (Continued)

voltage, V1 = 12 V, and the voltage-source current, I(V1) = 0.4 A, adhere to the passive convention. The power

$$V1 \cdot I(V1) = 12 \text{ V} \cdot 0.4 \text{ A} = 4.8 \text{ W}$$

is the power supplied *to* the voltage source *by* the rest of the circuit.
The current directed upward in the voltage source is given by

$$-I(V1) = -0.5 \text{ A}$$

The voltage source voltage, V1 = 12 V, and current, −I(V1) = 0.5 A, adhere to the active convention. The power supplied *by* the voltage source *to* the rest of the circuit is

$$V1 \cdot (-I(V1)) = 12 \text{ V} \cdot (-(-0.4 \text{ A})) = +4.8 \text{ W}$$

Often we shorten such statements and say that 4.8 W is the power supplied *by* the voltage source and −4.8 W is the power supplied *to* the voltage source.
Similarly, the power supplied by the current source is

$$-I1 \cdot V(b) = -0.5 \text{ A} \cdot (-4 \text{ V}) = 2 \text{ W}$$

The voltage and current of each of the resistors adhere to the passive convention. The power supplied to the 40 Ω resistor is

$$(V(a) - V(b)) \cdot (I(R1)) = (12 - (-4)) \cdot (0.4) = 6.4 \text{ W}$$

Similarly, 0.2 W are supplied to the 80 Ω resistor, 0.08 W are supplied to the 32 Ω resistor, and 0.12 W are supplied to the 48 Ω resistor.
Conservation of energy requires that the total power supplied by the two sources is equal to the total power supplied to the four resistors. Indeed that is the case.
The sources supply 4.8 W + 2 W = **6.8 W** and 5.4 W + 0.2 W + 0.08 W + 0.12 W = **6.8 W** are supplied to the resistors.
Step 6. Report the result.
The voltage source in the DC circuit in Figure 2.31 supplies 4.8 W and the current source supplies 2 W.
It is reasonable to ask how this example would have been different if we had made an error. Suppose we had inadvertently reversed the direction of the current source while drawing the LTspice schematic. The result would be the erroneous schematic as shown in Figure 2.36.

Example 2.7   (Continued)

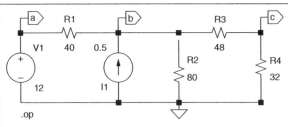

.op

Figure 2.36   The schematic corresponding to the circuit in Figure 2.31.

Of course we would fix this schematic if we noticed the error, but let us suppose that we do not notice the error and simulate the schematic to obtain the erroneous simulation results shown in Figure 2.37.

To specify the desired type of simulation, click on the Edit Tab and then select "Spice Analysis" to pop up the Edit Simulation Command dialog box. Select the "DC op pnt" tab, and then click the "OK" button. LTspice will generate a "spice directive" consisting of the command ".op." Position the spice directive on the schematic as shown in Figure 2.32.

Click on the Run LTspice toolbar icon, 🏃, to simulate the circuit.

The results of a DC analysis, labeled as the "Operating Point," display automatically as shown next:

Figure 2.38 shows the circuit from Figure 2.31 labeled with the Simulation Results from Figure 2.37.

Consider, for example, the voltage source in Figure 2.38. The voltage-source voltage, V1 = 12V, and the voltage-source current, I(V1) = 0.4A, adhere to the passive convention. The power

$$V1 \cdot I(V1) = 12\,V \cdot 0.4\,A = 4.8\,W$$

is the power supplied *to* the voltage source *by* the rest of the circuit.

```
                --- Operating Point ---

V(a):              12              voltage
V(b):              16              voltage
V(c):              6.4             voltage
I(I1):             0.5             device_current
I(R4):             0.2             device_current
I(R3):             0.2             device_current
I(R2):             0.2             device_current
I(R1):             -0.1            device_current
I(V1):             0.1             device_current
```

Figure 2.37   Simulation results: node voltages and device currents.

(Continued)

Example 2.7 (Continued)

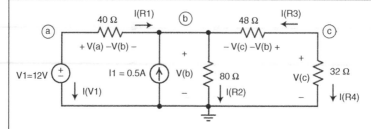

Figure 2.38 The circuit diagram from Figure 2.31, labeled with the simulation results from Figure 2.37.

The current directed upward in the voltage source is given by

$$-I(V1) = -0.5\,A$$

The voltage-source voltage, V1 = 12 V, and current, $-I(V1) = 0.5\,A$, adhere to the active convention. The power supplied *by* the voltage source *to* the rest of the circuit is

$$V1 \cdot (-I(V1)) = 12\,V \cdot (-(-0.4\,A)) = +4.8\,W$$

Often we shorten such statements and say that 4.8 W is the power supplied *by* the voltage source and −4.8 W is the power supplied *to* the voltage source.

Similarly, the power supplied by the current source is

$$-I1 \cdot V(b) = -0.5\,A \cdot (-4\,V) = 2\,W$$

The voltage and current of each of the resistors adhere to the passive convention. The power supplied to the 40 Ω resistor is

$$(V(a) - V(b)) \cdot (I(R1)) = (12 - (-4)) \cdot (0.4) = 6.4\,W$$

Similarly, 0.2 W are supplied to the 80 Ω resistor, 0.08 W are supplied to the 32 Ω resistor, and 0.12 W are supplied to the 48 Ω resistor.

Conservation of energy requires that the total power supplied by the two sources is equal to the total power supplied to the four resistors. Indeed that is the case.

The sources supply 4.8 W + 2 W = 6.8 W and 6.4 W + 0.2 W + 0.08 W + 0.12 W = 6.8 W are supplied to the resistors.

References

1 Svoboda, J.A. and Dorf, R.C. (2014). Dependent sources. In: *Introduction to Electric Circuits*, 9e, 33–37. Hoboken, NJ: Wiley.

2 Svoboda, J.A. and Dorf, R.C. (2014). Mesh equations. In: *Introduction to Electric Circuits*, 9e, 128–132. Hoboken, NJ: Wiley.

3 Svoboda, J.A. and Dorf, R.C. (2014). Equivalent circuit. In: *Introduction to Electric Circuits*, 9e, 63–73. Hoboken, NJ: Wiley.

4 Svoboda, J.A. and Dorf, R.C. (2014). Thevenin equivalent circuit. In: *Introduction to Electric Circuits*, 9e, 180–187. Hoboken, NJ: Wiley.

5 Svoboda, J.A. and Dorf, R.C. (2014). Superposition. In: *Introduction to Electric Circuits*, 9e, 176–171. Hoboken, NJ: Wiley.

Chapter 3

Variable DC Circuits

A "variable dc circuit" is a dc circuit in which the constant input is allowed to change. The input to a dc circuit has a constant value and so the values of all currents and voltages of the circuit, including mesh currents and node voltages, are constants. Consideration of variable dc circuits starts by asking the question "What would happen if the value of the input was changed from one constant value to another, different constant value?" Once the notion of changing the value of the constant input is considered, we can imagine varying the value of the input over a range of constant values.

One can imagine a laboratory experiment in which the value of the input to a circuit is set to a particular constant value and the resulting (constant) value of the circuit output is measured. Next, the value of the input is changed to a new constant value and the new value of the output is measured. This experiment is repeated for a range of constant input values. After all the data has been recorded, a plot can be constructed to display the relationship between the output of the circuit and the input.

LTspice simulates this type of experiment with its "DC Sweep" analysis.

3.1 DC Sweep

Example 3.1

Consider the circuit shown in Figure 3.1. The output voltage v_0 is related to the input voltage V_1 by the equation

$$v_0 = \frac{R_2 R_3}{R_1 R_2 + R_1 R_3 + R_2 R_3} V_1 + \frac{R_1 R_3}{R_1 R_2 + R_1 R_3 + R_2 R_3} V_2 \qquad (3.1)$$

(Continued)

LTspice® for Linear Circuits, First Edition. James A. Svoboda.
© 2023 John Wiley & Sons, Inc. Published 2023 by John Wiley & Sons, Inc.

Example 3.1 (Continued)

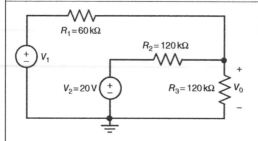

Figure 3.1 The circuit considered in Example 3.1.

Using the values given in Figure 3.1, we obtain

$$v_o = 0.5V_1 + 5 \tag{3.2}$$

Plotting v_o as a function of V_1 produces a straight line with a slope = 0.5 V/V and an intercept = 5 V. Let us use LTspice to simulate the circuit shown in Figure 3.1.

Step 1. Formulate a circuit analysis problem.

Plot the output voltage, v_o, of the circuit shown in Figure 3.1 as a function of the input voltage, V_1.

Step 2. Describe the circuit using an LTspice schematic.

Click on the New Schematic toolbar icon, 🗋, to start a new schematic. Place the component symbols and adjust the values of the circuit parameters. Place a ground symbol and wire the circuit. Label the output node to obtain the schematic shown in Figure 3.2.

Step 3. Simulate the circuit using LTspice.

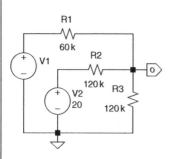

Figure 3.2 The circuit from Figure 3.1 represented as an LTspice schematic.

The voltage of the voltage source V1 in Figure 3.2 has not yet been specified. Let us make this voltage vary from −10 to 10 V in increments of 0.01 V. Select **Simulate/Edit Simulation Cmd** from the LTspice menu bar to pop up the "Edit Simulation Command" dialog box shown in Figure 3.3. Select the "DC sweep" tab and enter "V1" as the "Name of the 1st Source to Sweep." Specify values of the "Start Value," "Stop Value," and "Increment" as shown in Figure 3.3.

Example 3.1   (Continued)

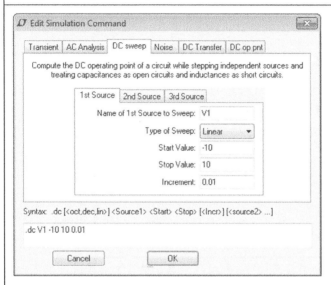

Figure 3.3 Setting up a DC sweep simulation.

Click the "OK" button to generate a "spice directive" consisting of the command ".dc V1 −10 10 0.01." Position the spice directive on the schematic as shown in Figure 3.4. Click on the Run LTspice toolbar icon, ✗, to simulate the circuit.

Step 4. Display the results of the simulation.

The results of a DC Sweep simulation display automatically as a blank plot. Click on the label of the output node to obtain a plot of V(o) versus V1. A label "V(o)" appears at the top of the border of the plot. Click on that label to obtain a cursor that can be positioned along the plot as shown in Figure 3.4. LTspice reports the position of the cursor in the Status Bar as shown in Figure 3.4. In the present case, x is the value of V1 and y is the value of V(o).

Step 5. Verify that the simulation results are correct.

Figure 3.4 shows the plot of V(o) versus V1 is a straight line as expected. The slope of that straight line is 0.5 V/V, as predicted in Eq. (3.2). Consequently

$$V(o) = 0.5V_1 + b \tag{3.3}$$

The cursor location in Figure 3.4 indicates that V(o) = 6 V when V1 = 1.97 V. Substituting into Eq. (3.3) gives

(Continued)

Example 3.1   (Continued)

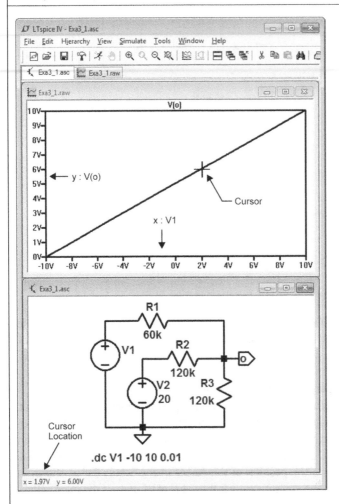

Figure 3.4   Simulation results.

$$6 = 0.5(1.97) + b \quad \Rightarrow \quad b = 5.015 \tag{3.4}$$

$$V(o) = 0.5V_1 + 5.015 \tag{3.5}$$

Equation (3.5) is quite similar to Eq. (3.2). The simulation results appear to be correct.

(Figure 3.5 illustrates an alternate view of the simulation results. Double-click on the label "V(o)" at the top of the border of the plot to obtain two cursors that can be positioned along the plot. A pop-up box appears to display the positions of both cursors as shown in Figure 3.5.)

Example 3.1   (Continued)

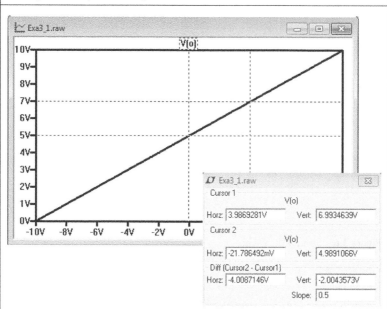

Figure 3.5   Alternate view of the simulation results.

Step 6. Report the answer to the circuit analysis problem.

Figure 3.5 shows a plot of the output voltage, V(o), of the circuit shown in Figure 3.1 as a function of the input voltage, V_1. This plot confirms that

$$v_o = 0.5V_1 + 5$$

Before leaving this example, let us see what happens if we use DC Sweep to vary the value of V2 as well as V1.

In Figure 3.6 the Edit Simulation Command dialog box is used to vary V2 from 12 to 24 V in increments of 8 V. LTspice sets V2 = 12 V and then sweeps V1 from −10 to 10 V in increments of 0.01 V. Next, LTspice sets V2 = 20 V and then sweeps V1 from −10 to 10 V in increments of 0.01 V. The corresponding simulation results are shown in Figure 3.7.

(Notice that

$$12 + 8 + 8 = 28 > 24 = \max V2$$

so V2 takes the values 12, 20, and 24 V in Figure 3.7.)

(*Continued*)

Example 3.1   (Continued)

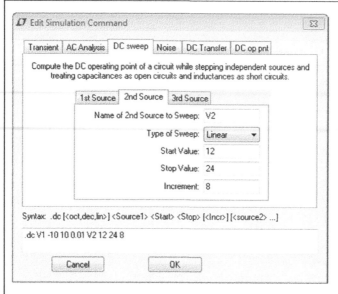

Figure 3.6   Second DC sweep.

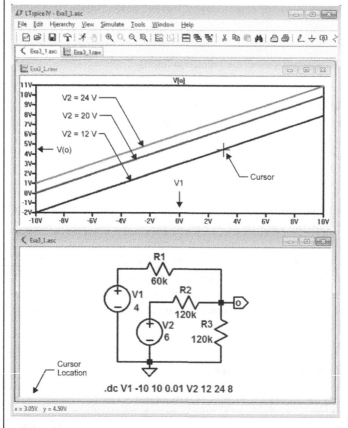

Figure 3.7   Simulation results.

3.2 Global Parameters

The circuit shown in Figure 3.8 is called a voltage divider.

The output voltage v_o is related to the input voltage v_i by the equation

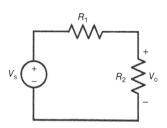

$$v_0 = \frac{R_2}{R_1 + R_2} V_s \qquad (3.6)$$

The gain of the voltage divider is

Figure 3.8 A voltage divider.

$$\text{gain} = \frac{v_0}{V_s} = \frac{R_2}{R_1 + R_2} \qquad (3.7)$$

Plotting v_0 as a function of V_s produces a straight line that passes through the origin. The value of the slope of this line is equal to the value of the gain of the voltage divider.

The input resistance of the voltage divider is

$$R_i = R_1 + R_2 \qquad (3.8)$$

The input resistance of the voltage divider is significant because it determines the value of the power, p_s, that the voltage source must supply.

$$p_s = \frac{V_s^2}{R_1 + R_2} \qquad (3.9)$$

It is reasonable to describe a voltage divider by specifying required values of input resistance and then gain and to design the voltage divider by determining values of R_1 and R_2 that produce the specified input resistance and gain. Solving Eqs. (3.7) and (3.8) for R_1 and R_2 gives

$$R_1 = (1 - \text{gain}) \times R_i \qquad (3.10)$$

and

$$R_2 = \text{gain} \times R_i \qquad (3.11)$$

An LTspice feature called "Global Parameters" makes it possible to specify values of the input resistance and gain in a schematic and then use these values to calculate corresponding values for R_1 and R_2.

Example 3.2

Use LTspice to design a voltage divider having specified values of gain and input resistance.

Step 1. Formulate a circuit analysis problem.

Design a voltage divider to have gain = 0.8 V/V and input resistance = 100 kΩ. Plot the resistances R_1 and R_2 of voltage dividers having input resistance R_i = 100 kΩ and gains ranging from 0.05 to 0.95.

Step 2. Describe the circuit using an LTspice schematic.

Click on the New Schematic toolbar icon, [𝄃], to start a new schematic. Place the component symbols, place a ground symbol, and wire the circuit. Label the output node to obtain the schematic shown in Figure 3.10a.

Next, we will add two global parameters to represent the input resistance and gain. Then, we will modify the properties of the resistors R_1 and R_2 so that their resistances depend on the global parameters.

Click on the SPICE directive icon, .op, at the right end of the LTspice toolbar to open the dialog box shown in Figure 3.9. Enter

$$.param\, gain = 0.8, Ri = 100k$$

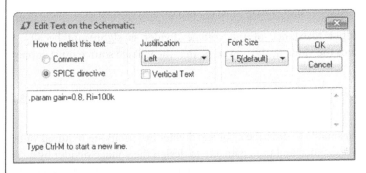

Figure 3.9   SPICE directive dialog box.

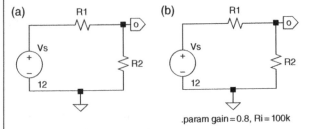

Figure 3.10   Schematic (a) before and (b) after adding the SPICE directive.

Example 3.2 (Continued)

in the text box as shown in Figure 3.9. Place the SPICE direction on the schematic as shown in Figure 3.10b.

Now we can use the parameters "gain" and "Ri" to specify the resistances R1 and R2. Right-click on resistor R1 and enter {gain*Ri} as shown in Figure 3.11a. The braces in {gain*Ri} that the value of the resistance is obtained by evaluating the expression enclosed by the braces:

$$R2 = gain*Ri = 0.8*100k = 80k\Omega$$

Specify the value of R2 as {(1−gain)*Ri} and label node i to obtain the schematic in Figure 3.11b.

Step 3. Simulate the circuit using LTspice.

Click on the Edit Tab and then select "Spice Analysis" to pop up the Edit Simulation Command dialog box. Select the "DC op pnt" tab and then click the "OK" button. LTspice will generate a "spice directive" consisting of the command ".op." Position the spice directive on the schematic as desired.

Click on the Run LTspice toolbar icon, ✗, to simulate the circuit.

Step 4. Display the results of the simulation.

The results of a dc analysis display automatically and are labeled as the "Operating Point" as shown in Figure 3.12. The operating point consists of node voltages in Volts and the circuit element currents in Amps.

Step 5. Verify that the simulation results are correct.

We can use Ohm's and Kirchhoff's laws to determine the values of R_1 and R_2 from the simulations results:

(a)   (b)

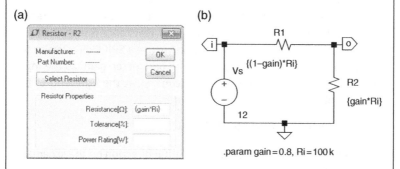

Figure 3.11 (a) Specifying R2 and (b) the schematic after specifying both R1 and R2.

(*Continued*)

Example 3.2   (Continued)

```
    --- Operating Point ---
V(i):     12          voltage
V(o):     9.6         voltage
I(R2):    0.00012     device_current
I(R1):    0.00012     device_current
I(Vs):    -0.00012    device_current
```

Figure 3.12   Simulation results.

$$R_1 = \frac{V(i) - V(o)}{I(R1)} = \frac{12 - 9.6}{0.00012} = 20\,k\Omega \text{ and}$$

$$R_2 = \frac{V(o)}{I(R2)} = \frac{9.6}{0.00012} = 80\,k\Omega$$

The input resistance is

$$R_i = R_1 + R_2 = 20\,k\Omega + 80\,k\Omega = 100\,k\Omega$$

Next, we can determine the value of the gain from the simulation results. Noticing that the output is $V(o)$ = 9.6 V when the input is $V(i)$ = 12 V, the gain is given by

$$gain = \frac{V(o)}{V(i)} = \frac{9.6}{12} = 0.8\,V/V$$

Both the input resistance and the gain have their specified values. The simulation results are correct.

Step 6. Report the answer to the circuit analysis problem.

We can use LTspice to plot the values of the voltage divider resistances R_1 and R_2 corresponding to an input resistance R_i = 100 kΩ and gains ranging from 0.05 to 0.95. To do so, we use a SPICE directive to tell LTspice to perform a DC Sweep using the parameter "gain" as the input variable. Click on the SPICE directive icon, .op, at the right end of the LTspice toolbar to open a dialog box. Enter

.step param gain 0.05 0.95 0.01

in the text box and click OK. Place the SPICE direction on the schematic as shown in Figure 3.13.

Example 3.2 (Continued)

Click on the Run LTspice toolbar icon, ✈, to simulate the circuit. A blank plot with display. Click on the label for node o to obtain a plot of V(o) versus the gain. The label "V(o)" appears at the top of this plot. Right-click on that label to edit that label. Change V(o) to V(o)/I(R2), which is equal to R2.

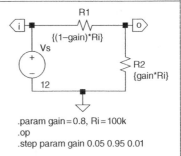

.param gain = 0.8, Ri = 100k
.op
.step param gain 0.05 0.95 0.01

Figure 3.13 A voltage divider.

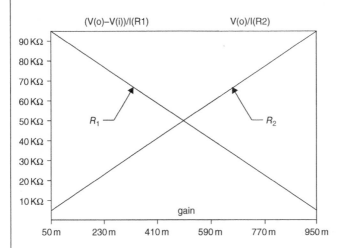

Figure 3.14 Resistance values versus the gain of the voltage divider.

Click on the label for node o a second time. Once again, the label "V(o)" appears at the top of this plot. Right-click on "V(o)" to edit that label to change it to (V(o)−V(i))/I(R1), which is equal to R1.

Right-click on an empty space just above the plot to obtain a dialog box that can be used to adjust the range and increment of each axis. Figure 3.14 shows the final plot.

Example 3.3

Use LTspice to design a temperature-dependent voltage divider.

The resistor labeled R_T in the voltage divider shown in Figure 3.15a is a thermistor. A thermistor, shown in Figure 3.15b, is a semiconductor device that exhibits a temperature-dependent resistance described by the equation

$$R_T = R_0\, e^{\beta\left(\frac{1}{T}-\frac{1}{T_0}\right)} \tag{3.12}$$

where R_T is the thermistor resistance, in Ω, at temperature T in °K and R_0 is the thermistor resistance, in Ω, at temperature T_0 in °K. β is a constant that depends on the material used to manufacture the thermistor.

Suppose the thermistor shown in Figure 3.15b has $\beta = 3200$ K and $R = 200\,\Omega$ at $T = 298$ K $= 22$ °C. The resistance of this thermistor is given by

$$R_T = 200\, e^{3200\left(\frac{1}{T}-\frac{1}{298}\right)} \tag{3.13}$$

The output voltage of the voltage divider in Figure 3.15a is given by

$$v_o = \frac{9R_T}{50 + R_T} \tag{3.14}$$

Equations (3.13) and (3.14) enable us to determine the thermistor temperature from the output voltage of the voltage divider.

Step 1. Formulate a circuit analysis problem.

a) Determine the value of the thermistor voltage, v_o, when the thermistor temperature is 77 °F.

b) Determine the value of the thermistor temperature in °F indicated by $v_o = 6$ V.

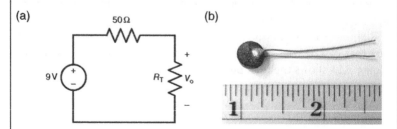

(a)   50 Ω   (b)

9 V R_T v_o

Figure 3.15 (a) A voltage divider with a thermistor and (b) the thermistor.

Example 3.3   (Continued)

Step 2. Describe the circuit using an LTspice schematic.

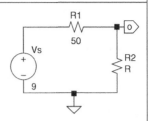

Click on the New Schematic toolbar icon, 🖉, to start a new schematic. Place the component symbols and adjust the values of the circuit parameters. Place a ground symbol and wire the circuit. Label the output node to obtain the schematic shown in Figure 3.16.

Figure 3.16   A voltage divider schematic.

Resistor R2 is a thermistor having resistance R_T given by Eq. (3.13). Let us add Eq. (3.13) to the schematic. Click on the SPICE directive icon, ·op, at the right end of the LTspice toolbar to open a dialog box. Enter

$$.param\ beta = 3200,\ Ro = 200,\ To = 273 + 25,\ T = 273 + temp$$
$$.param\ RT = Ro^*exp\big(beta^*\big(1/T - 1/To\big)\big)$$
$$.step\ temp\ 0\ 100\ 0.5$$

(a)

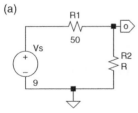

.param beta = 3200, Ro = 200, To = 273 + 25, T = 273 + temp
.param RT = Ro*exp(beta*(1/T − 1/To))
.step temp 0 100 0.5

(b)

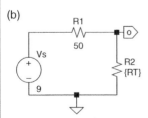

.param beta = 3200, Ro = 200, To = 273 + 25, T = 273 + temp
.param RT = Ro*exp(beta*(1/T − 1/To))
.step temp 0 100 0.5
.op

Figure 3.17   A voltage divider with a thermistor represented as an LTspice schematic.

(*Continued*)

Example 3.3 (Continued)

in the dialog box, click OK, and place the SPICE directive on the schematic as shown in Figure 3.17a. The SPICE directive defines parameters beta, Ro, To, and T. In contrast, "temp" is a parameter defined by LTspice to be the ambient temperature in °C. The SPICE directive specifies that the ambient temperature ranges from 0 to 100 °C in 0.01 °C increments. Right-click on the thermistor and set the thermistor resistance to {RT} as shown in Figure 3.17b.

Step 3. Simulate the circuit using LTspice.

Click on the Edit Tab and then select "Spice Analysis" to pop up the Edit Simulation Command dialog box. Select the "DC op pnt" tab and then click the "OK" button. LTspice will generate a "spice directive" consisting of the command ".op." Position the spice directive on the schematic as desired.

Click on the Run LTspice toolbar icon, 🏃, to simulate the circuit.

Step 4. Display the results of the simulation.

The results of a DC Sweep simulation display automatically as a blank plot. Click on the label of the output node to obtain a plot of V(o) versus temp. A label "V(o)" appears at the top of the border of the plot. Click on that label twice to obtain two cursors that can be positioned along the plot as shown in Figure 3.18. These cursors indicate that

$$V(o) = 7.1926432\,V \text{ when temp} = 25.08056\,°C$$
$$\text{and that temp} = 45.51495\,°C \text{ when } V(o) = 6.0030309\,V.$$

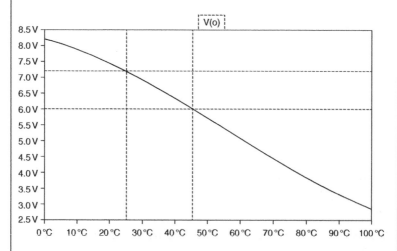

Figure 3.18 Plot of the voltage divider output voltage versus the thermistor temperature.

Example 3.3   (Continued)

Step 5. Verify that the simulation results are correct.

Let us consider each of the two situations identified by the cursors in Figure 3.18. First, suppose V(o) = 7.19643 V. Use Eq. (3.14) that the corresponding thermistor resistance is

$$R_T = \frac{50v_o}{9 - v_o} = \frac{50(7.19643)}{9 - 7.19643} = 198.98 \,\Omega \qquad (3.15)$$

then use Eq. (3.13) to determine the corresponding thermistor temperature.

$$T = \frac{1}{\dfrac{\ln\left(\dfrac{R_T}{R_0}\right)}{\beta} - \dfrac{1}{T_0}} = \frac{1}{\dfrac{\ln\left(\dfrac{198.98}{200}\right)}{3200} - \dfrac{1}{298}} = 298.14\,°\text{K} = 25.14\,°\text{C} \qquad (3.16)$$

Next, suppose T = temp = 45.515 °C = 318.515 °K. Use Eq. (3.13) that the corresponding thermistor resistance is

$$R_T = 200e^{3200\left(\frac{1}{318.515} - \frac{1}{298}\right)} = 100.15 \,\Omega$$

then use Eq. (3.14) to determine the value of the output voltage of the voltage divider.

$$v_o = \frac{9R_T}{50 + R_T} = \frac{9(101.15)}{50 + 101.15} = 6.03 \,\text{V}$$

In both cases the simulation results are consistent with Eqs. (3.13) and (3.14) so we are confident that the simulation is correct.

Step 6. Report the answer to the circuit analysis problem.

a) The output voltage of the voltage divider is v_o = 6.0 V when the thermistor temperature is 77 °F = 45.515 °C.

b) The thermistor temperature is 25.14 °C when the output voltage of the voltage divider is v_o = 7.19643 V.

Example 3.4

Use LTspice to analyze a circuit containing ideal op amps.

Step 1. Formulate a circuit analysis problem.

The output of the circuit shown in Figure 3.19 is the voltage v_o at the right of the circuit. The input is the voltage-source voltage v_s. Express v_o as a function of v_s.

Step 2. Describe the circuit using an LTspice schematic.

The circuit in Figure 3.19 contains two ideal op amps [1]. We can implement an ideal op in LTspice using the component named UniversalOpAmp2 shown in Figure 3.20a. The UniversalOpAmp2 has five terminals: three terminals of the ideal op amp plus two additional terminals for power supplies. Power supplies for the UniversalOpAmp2 used as an ideal op amp consist of 15-V voltage sources and Bi-Direct terminals as shown in Figure 3.20b. (Click on the tool bar icon ⓐ and select "Bi-Direct" from the dropdown menu named "Port-Type.") The LTspice schematic shown in Figure 3.20c represents an ideal op amp that saturates [2] at ±15 V.

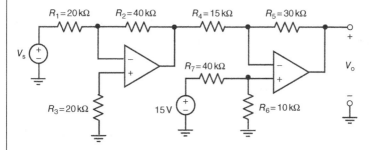

Figure 3.19 Circuit containing two op amps.

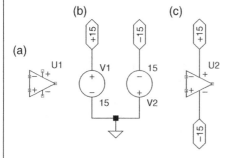

Figure 3.20 (a) UniversalOpAmp2; (b) power supplies; (c) ideal Op Amp.

Example 3.4 (Continued)

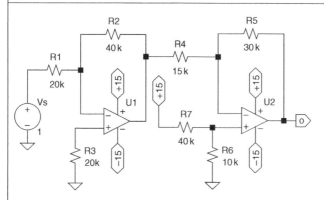

Figure 3.21   LTspice schematic corresponding to the circuit in Figure 3.19.

Figure 3.21 shows an LTspice schematic corresponding to the circuit in Figure 3.19.

Step 3. Simulate the circuit using LTspice.

Select **Simulate/Edit Simulation Cmd** from the LTspice menu bar to pop up the "New Simulation" dialog box. Specify a "DC Sweep" of source "Vs" with a Start Value of −8, a Stop Value of 4, and an Increment of 0.05. Click the "OK" button to generate a "spice directive" consisting of the command ".dc Vs −8 4 0.05." Position the spice directive on the schematic as desired. Click on the Run LTspice toolbar icon, 🦌, to simulate the circuit.

Step 4. Display the results of the simulation.

The results of a DC Sweep simulation display automatically as a blank plot. Click on the label of the output node to obtain a plot of V(o) versus Vs. A label "V(o)" appears at the top of the border of the plot. Click on that label twice to obtain two cursors that can be positioned along the plot as shown in Figure 3.22.

The plot in Figure 3.22 is described by the following equation:

$$V(o) = \begin{cases} -15 & \text{when} & Vs < -6V \\ 4Vs + 9 & \text{when} & -6V < Vs < 1.5V \\ +15 & \text{when} & Vs > 1.5V \end{cases} \tag{3.17}$$

(Continued)

Example 3.4   (Continued)

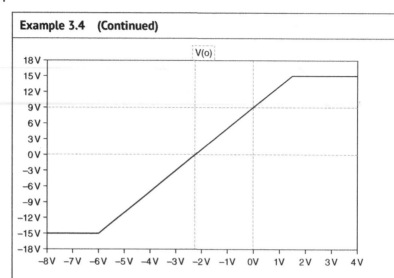

Figure 3.22   Simulation results.

Step 5. Verify that the simulation results are correct.

Figure 3.23 shows the consequences of assuming that the op amps in the circuit shown in Figure 3.19 are ideal op amps. In particular, the input currents of both op amps are labeled as "0 A" and the input voltages of each op amp are labeled as equal voltages.

Consider the case in which the input voltage, v_s, is between −6 and 6 V. Apply Kirchhoff's Current Law (KCL) at the inverting input of the left op amp to obtain

$$v_1 = -\frac{R_2}{R_1} v_s \tag{3.18}$$

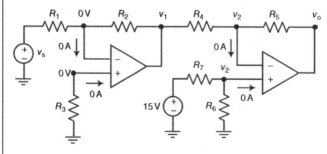

Figure 3.23   Circuit containing two ideal op amps.

Example 3.4 (Continued)

Apply KCL at the noninverting input of the right op amp to obtain

$$v_2 = \frac{R_6}{R_6 + R_7} 15 \tag{3.19}$$

Apply KCL at the inverting input of the right op amp to obtain

$$v_o = \left(1 + \frac{R_5}{R_4}\right) v_2 - \frac{R_5}{R_4} v_1 \tag{3.20}$$

Substitute Eqs. (3.18) and (3.19) into Eq. (3.20) to obtain

$$v_o = \frac{R_5 R_2}{R_4 R_1} v_s + \left(1 + \frac{R_5}{R_4}\right) \frac{R_6}{R_6 + R_7} 15 \tag{3.21}$$

Finally, substitute the resistance values from Figure 3.19 into Eq. (3.20) to obtain

$$v_o = 4v_s + 9 \tag{3.22}$$

Equation (3.22) agrees with Eq. (3.17) so we are confident that the simulation results are correct.

Step 6. Report the answer to the circuit analysis problem.

The output, v_o, of the circuit shown in Figure 3.19 is the input, v_s, by the equation

$$v_o = \begin{cases} -15 & \text{when} \quad v_s < -6\,\text{V} \\ 4v_s + 9 & \text{when} \quad -6\,\text{V} < v_s < 1.5 \\ +15 & \text{when} \quad v_s > 1.5\,\text{V} \end{cases}$$

References

1 Svoboda, J.A. and Dorf, R. *Introduction to Electric Circuits*, 9e, 221–223. Hoboken, NJ: Wiley.

2 Svoboda, J.A. and Dorf, R. *Introduction to Electric Circuits*, 9e, 245–247. Hoboken, NJ: Wiley.

Chapter 4

Capacitors and Inductors

4.1 Capacitors

Capacitors are circuit elements consisting of two isolated conductors. Often these conductors are visualized as parallel plates. Capacitors in an electric circuit store charge. The capacitance, C, of the capacitor is the ratio of the amount of charge, q, stored by the capacitor to the voltage, v, across the capacitor:

$$C = \frac{q}{v} \quad \Rightarrow \quad q = Cv \tag{4.1}$$

The capacitor current, i, is given by the derivative of the charge stored by the capacitor:

$$i = \frac{dq}{dt} = C\frac{dv}{dt} \tag{4.2}$$

Integrating both sides of Eq. (4.2) gives

$$\int_{-\infty}^{t} i(\tau)\,d\tau = Cv \tag{4.3}$$

so

$$v(t) = \frac{1}{C}\int_{-\infty}^{t} i(\tau)\,d\tau \tag{4.4}$$

LTspice simulations begin at time $t = 0$ so it is convenient to write Eq. (4.4) as

$$v(t) = \frac{1}{C}\int_{-\infty}^{0} i(\tau)\,d\tau + \frac{1}{C}\int_{0}^{t} i(\tau)\,d\tau = v(0) + \frac{1}{C}\int_{0}^{t} i(\tau)\,d\tau \tag{4.5}$$

LTspice® for Linear Circuits, First Edition. James A. Svoboda.
© 2023 John Wiley & Sons, Inc. Published 2023 by John Wiley & Sons, Inc.

The voltage $v(0)$ is called the initial condition of the capacitor. (The capacitor voltage and current in Eq. (4.5), v and i, adhere to the passive convention [1]. That is, the capacitor current is directed from the + sign toward the − sign of the polarity of the capacitor voltage.)

Example 4.1

Figure 4.1 shows circuit consisting of a capacitor and a voltage source. The input to this circuit is the voltage-source voltage, V_s. The capacitor voltage $v_c(t)$ is equal to the voltage-source voltage and the capacitor current is the output of the circuit.

The voltage-source voltage, V_s, switches back and forth between two values, 1 and 11 V. We see that the voltage remains at one value for 100 ms before switching to the other value. The transitions between 1 and 11 V appear to be instantaneous, but that cannot be true. Equation (4.2) indicates that an infinite current is necessary if the capacitor voltage is to change instantaneously from 1 to 11 V or vice versa. That is not possible and sounds dangerous. More likely, the transition occurs quickly enough that it appears instantaneous on the scale used to plot the voltage-source voltage in Figure 4.1a.

For the sake of simplicity, let us assume that the transitions of V_s between 1 and 11 V occur as linear functions of time as shown in Figure 4.2. This waveform is called a "pulse function" in LTspice and can be used to specify the voltage of a voltage source or current of a current source. Figure 4.3 illustrates the terminology used to specify a pulse in LTspice.

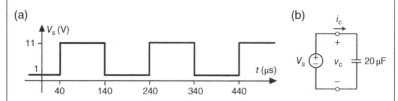

Figure 4.1 The circuit considered in Example 4.1 (b) and the voltage of the source (a).

Example 4.1   (Continued)

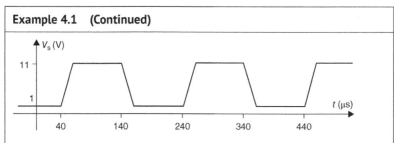

Figure 4.2   Source voltage from Figure 4.1a as modified to avoid instantaneous changes in voltage.

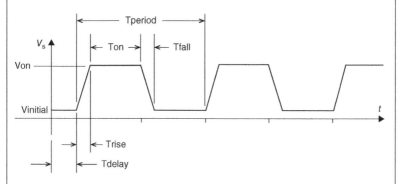

Figure 4.3   A voltage pulse labeled using LTspice terminology.

Step 1. Formulate a circuit analysis problem.

Determine the value of the maximum capacitor current in Figure 4.1b when the voltage-source voltage is the pulse function shown in Figure 4.2 with Tfall = Trise = 20 μs.

Step 2. Describe the circuit using an LTspice schematic.

Click on the New Schematic toolbar icon, 🗗, to start a new schematic. Place symbols representing the voltage source and capacitor.

Right-click on the voltage source, then click the "Advanced" button in the "Voltage Source – V1" dialog box, and then specify a pulse voltage as shown in Figure 4.4.

Next, right-click on the capacitor and specify the value of the capacitance. Place a ground symbol and wire the circuit. Add a node label as shown in Figure 4.5.

Step 3. Simulate the circuit using LTspice.

Select **Simulate/Edit Simulation Cmd** from the LTspice menu bar to pop up the "New Simulation" dialog box. Specify a "Transient" simulation with a "Stop time" equal to 500 μs. LTspice will generate a "spice directive"

(Continued)

Example 4.1   (Continued)

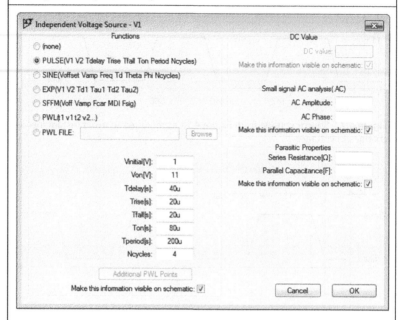

Figure 4.4   AC voltage source dialog box.

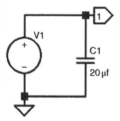

Figure 4.5   Example 4.1 schematic.

PULSE(1 11 40 u 20 u 20 u 80 u 200 u 4)
.tran 500 u

consisting of the command ".tran 500u." Position the spice directive on the schematic as shown in Figure 4.5.

Click on the LTspice toolbar Run icon, ⚡, to simulate the circuit.

Step 4. Display the results of the simulation.

Step 5. Verify that the simulation results are correct.

In Figure 4.6a we see that the capacitor voltage increases from 1 to 11V in Trise = 20μs. Equation (4.2) indicates that the corresponding capacitor current is

$$i(t) = C_1 \frac{dv}{dt} = 20\,\mu F \frac{10\,V}{20\,\mu s} = 10\,A$$

Example 4.1   (Continued)

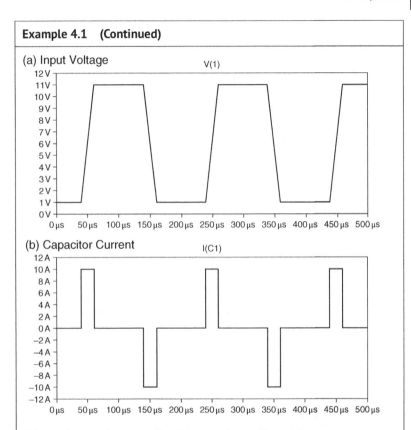

Figure 4.6   Simulation results: (a) input voltage; (b) capacitor current.

We see in Figure 4.6b that the capacitor current is indeed 10 A while capacitor voltage increases from 1 to 11 V as shown in Figure 4.6a. Also, we see in Figure 4.6b that the capacitor voltage decreases from 11 to 1 V in 20 µs requiring a capacitor current

$$i(t) = C_1 \frac{dv}{dt} = 20\,\mu F \frac{-10\,V}{20\,\mu s} = -10\,A$$

We see in Figure 4.6b that the capacitor current is indeed −10 A while capacitor voltage decreases from 11 to 1 V.

We also notice that the capacitor current is 0 A when the capacitor voltage is constant, whether at 11 V or at 1 V. The simulation results appear to be correct.

Step 6. Report the answer to the circuit analysis problem.

The size of the maximum capacitor current in Figure 4.1b is 10 A when the voltage-source voltage is the pulse function shown in Figure 4.2 with Tfall = Trise = 20 µs.

(Continued)

Example 4.2

Figure 4.7a shows a circuit consisting of two capacitors and a switch that is initially open and then closes at time $t = 1\,\mu s$. Ideally, an open switch acts like an open circuit and a closed switch acts like a short circuit. A more accurate model of a switch represents an open switch as a large resistance and a closed switch as a small resistance. (The model "SW" incorporated into LTspice represents an open switch as a resistor $R_{off} = 1\,M\Omega$ and a closed switch by a resistor $R_{on} = 0.1\,\Omega$.)

Figure 4.7b shows the circuit after time $t = 1\,\mu s$ with the resistor R_{on} representing the closed switch. The values of the charge stored by the capacitors are given by

$$q_1 = \frac{v_1}{C_1} = \frac{10\,V}{2 \times 10^{-6}\,F} = 2 \times 10^{-6}\,\text{Coulombs} \quad \text{and}$$

$$q_2 = \frac{v_2}{C_2} = \frac{4\,V}{10 \times 10^{-6}\,F} = 0.4 \times 10^{-6}\,\text{Coulombs}$$

The switch current, i, depends on the capacitor voltages as

$$i = \frac{v_1 - v_2}{R_{on}}$$

This current represents a flow of charge through the resistor from capacitor C_1 to capacitor C_2. After the switch closes and while $v_1 > v_2$, we expect the charge stored by capacitor C_1 to decrease and the charge stored by capacitor C_2 to increase. Consequently, voltage v_2 will increase and voltage v_1 will decrease. The switch current will decrease from its initial value of $\dfrac{10\,V - 4\,V}{0.1\,\Omega} = 60\,A$ at $t = 1\,\mu s$ to its final value of 0 A when v_2 becomes equal to v_1.

Next, we use LTspice to determine the voltages across the capacitors in the circuit shown in Figure 4.7b after the switch closes. (When the

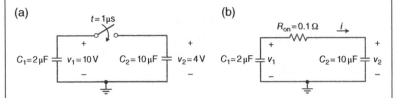

(a)

(b)

Figure 4.7 (a) The circuit considered in Example 4.2 and (b) the circuit after replacing the switch with a model of a closed switch.

Example 4.2   (Continued)

switch is open, the switch current is negligibly small and is ignored in this example.)

Step 1. Formulate a circuit analysis problem.

Analyze the circuit shown in Figure 4.7a to plot the capacitor voltages after the switch closes.

Step 2. Describe the circuit using an LTspice schematic.

Click on the New Schematic toolbar icon, 🖵, to start a new schematic. Place symbols representing the capacitors. Specify the capacitance values. Label the top node of C1 as node 1 and the top node of C2 as node 2. Label the bottom nodes of both capacitors as the ground node as shown in Figure 4.8a.

Click on the Component icon, ⅁, on the toolbar to pop up the Select Component Symbol dialog box. Start typing "switch" in the search box. LTspice will find the symbol for a switch as soon as "sw" has been typed. Click the "OK" button on the Select Component Symbol dialog box. A switch symbol will appear on the schematic. Rotate the switch symbol, e.g. by typing <CNTL>R three times, and insert the switch symbol into the schematic as shown in Figure 4.8b.

The switch requires a "spice directive" called a "model statement." Click on the SPICE directive icon, ·op, at the right end of the LTspice toolbar to open a dialog box. Select "spice directive" and enter

.model SW SW( )

in the text box. Position the spice directive on the schematic as shown in Figure 4.8c.

The switch symbol has terminals labeled "−" and "+." Ground the "−" terminal and connect a grounded voltage source to the "+" terminal as shown in Figure 4.8d.

Right-click on the voltage source and click the "Advanced" button in the "Voltage Source – V1" dialog box, and then specify a pulse voltage having Vinitial[V] = −1, Von[V] = 1, and Tdelay[s] = 1 µs.

Right-click on the voltage source and click the "Advanced" button in the "Voltage Source – V1" dialog box, and then specify the amplitude and phase of the voltage-source voltage as shown in Figure 4.4.

Step 3. Simulate the circuit using LTspice.

Select **Simulate/Edit Simulation Cmd** from the LTspice menu bar to pop up the "New Simulation" dialog box. Specify a "Transient" simulation

(Continued)

Example 4.2   (Continued)

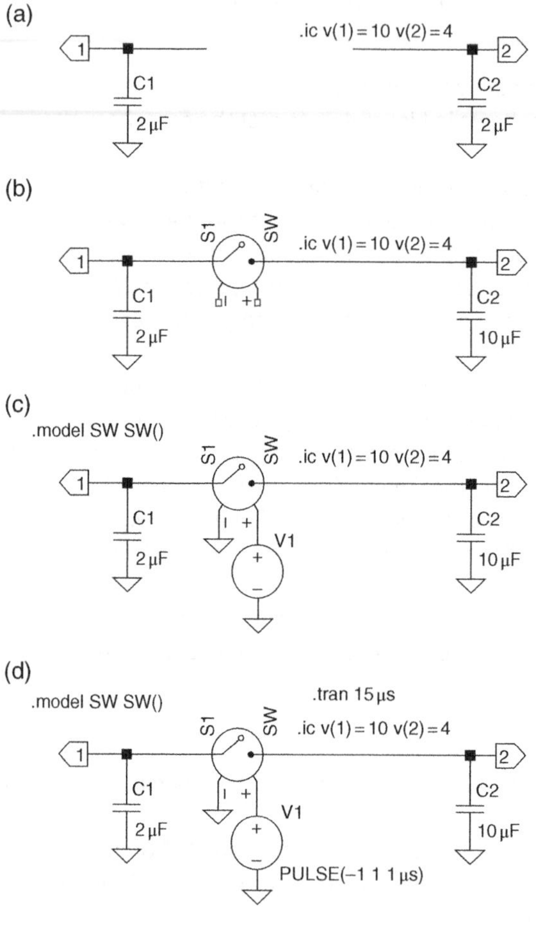

(a)

.ic v(1) = 10 v(2) = 4

〈1│
C1
2 μF

C2
│2〉
2 μF

(b)

S1   SW

〈1│
C1
2 μF

.ic v(1) = 10 v(2) = 4

C2
│2〉
10 μF

(c)

.model SW SW()

S1   SW   .ic v(1) = 10 v(2) = 4

〈1│
C1
2 μF

V1

C2
│2〉
10 μF

(d)

.model SW SW()

.tran 15 μs

S1   SW   .ic v(1) = 10 v(2) = 4

〈1│
C1
2 μF

V1

PULSE(−1 1 1 μs)

C2
│2〉
10 μF

Figure 4.8   LTspice schematic corresponding to the circuit of Figure 4.7.

with a "Stop time" equal to 15 μs. LTspice will generate a "spice directive" consisting of the command ".tran 15 μs." Position the spice directive on the schematic as shown in Figure 4.8d.

Click on the LTspice toolbar Run icon, 🏃, to simulate the circuit.

Step 4. Display the results of the simulation.

An empty plot displays automatically. Click on the labels of the node 1 and node 2 of the schematic to display the plot of the voltages at node 1

Example 4.2 (Continued)

and node 2 shown in Figure 4.9. (The capacitor voltages are equal to the node voltages at nodes 1 and 2.)

Step 5. Verify that the simulation results are correct.

Charge is conserved in an electric circuit [2]. In this example, that means that the total charge stored by the capacitors after the capacitor voltages have become equal is equal to the total charge stored by the capacitors before the switch opens. Using Eq. (4.1) and the initial capacitor voltages from Figure 4.7a, we determine that the charge stored on the capacitors before the switch opens to be

$$\text{Total charge before the switch opens} = C_1 v_1 + C_2 v_2$$
$$= (2\,\mu F)(10\ V) + (10\,\mu F)(4\ V) = 60\ \mu C$$

and

$$\text{Total charge after the capacitor voltages have become equal}$$
$$= (C_1 + C_2)v_f = (12\,\mu F)v_f$$

Charge is conserved so

$$60\,\mu C = (12\,\mu F)v_f \quad \Rightarrow \quad v_f = \frac{60\,\mu C}{12\,\mu F} = 5\ V$$

Step 6. Report the answer to the circuit analysis problem.

Figure 4.9 shows the voltages across the capacitors in the circuit shown in Figure 4.7 after the switch closes. After about 10 μs the capacitor voltages are essentially equal with a value of 5 V.

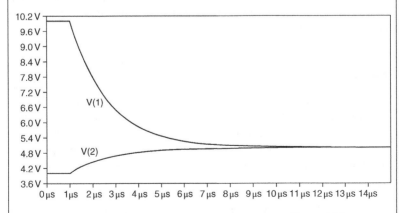

Figure 4.9 The node voltages of the schematic shown in Figure 4.8d.

4.2 Inductors

An inductor is a circuit element that stores energy in a magnetic field. An inductor can be constructed by winding a coil of wire around a magnetic core [3]. A current, i, in the coil causes a voltage, v, across the coil. The voltage across the coil is proportional to the rate of change of the coil current, that is

$$v = L\frac{di}{dt} \tag{4.6}$$

where the constant of proportionality, L, is called the **inductance** of the inductor. The unit of inductance is called henry (H).

Integrating both sides of Eq. (4.6) gives

$$\int_{-\infty}^{t} v(\tau)d\tau = Li \tag{4.7}$$

so

$$i(t) = \frac{1}{L}\int_{-\infty}^{t} v(\tau)d\tau \tag{4.8}$$

LTspice simulations begin at time $t = 0$ so it is convenient to write Eq. (4.8) as

$$i(t) = \frac{1}{L}\int_{-\infty}^{0} v(\tau)d\tau + \frac{1}{L}\int_{0}^{t} v(\tau)d\tau = i(0) + \frac{1}{L}\int_{0}^{t} v(\tau)d\tau \tag{4.9}$$

The current $i(0)$ is called the initial condition of the inductor. (The inductor voltage and current, v and i, in Eq. (4.10) adhere to the passive convention [1]. That is, the inductor current is directed from the $+$ sign toward the $-$ sign of the polarity of the inductor voltage.)

Example 4.3

The current of the current source in the circuit shown in Figure 4.10a is a piecewise linear function of time. In other words, it consists of a series of straight line segments. Also, consecutive straight line segments share end points. It is convenient to represent these straight line segments by a list of end points. For example, the current $i_s(t)$ in Figure 4.10b can be represented by

$$(0,0),(2,8),(4,0),(6,-2),(15,-2),(16,0),(20,0)$$

Example 4.3 **(Continued)**

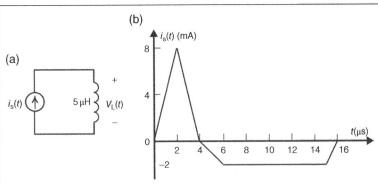

Figure 4.10 A circuit (a) with a piecewise linear input (b).

Each of the points is represented by an ordered pair, $(t_k, i_k(t_k))$, where t_k is the time in μs and $i_k(t_k)$ is the corresponding current in mA.

Step 1. Formulate a circuit analysis problem.

Simulate the circuit shown in Figure 4.10 to determine the inductor voltage, $v_L(t)$. Plot the inductor voltage, $v_L(t)$ versus t. Assume $i_s(t) = 0$ for $t \leq 0$.

Step 2. Describe the circuit using an LTspice schematic.

Click on the New Schematic toolbar icon, ⟨⟩, to start a new schematic. Place symbols representing the current source and inductor.

Right-click on the current source and click the "Advanced" button in the "Current Source – V1" dialog box, select a PWL (Piecewise Linear) current function and enter the list of end points given earlier.

Specify the value of the inductance. Place a ground symbol and wire the circuit. Add a node label to obtain the schematic shown in Figure 4.11.

Step 3. Simulate the circuit using LTspice.

Select **Simulate/Edit Simulation Cmd** from the LTspice menu bar to pop up the "New Simulation" dialog box. Specify a "Transient" simulation, and then specify a "Stop Time" of 20 μs to request a simulation that begins at time 0 and ends at 20 μs. Click "OK" to close the "New Simulation" dialog box.

Click on the Run LTspice toolbar icon, ⚲, to simulate the circuit.

(Continued)

Example 4.3   (Continued)

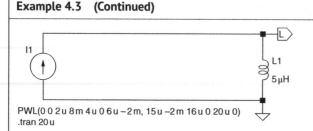

PWL(0 0 2u 8m 4u 0 6u −2m, 15u −2m 16u 0 20u 0)
.tran 20u

Figure 4.11   LTspice schematic for the circuit shown in Figure 4.10.

Step 4. Display the results of the simulation.

The results of a "Transient" simulation display automatically as a blank plot. Click on the inductor to obtain a plot of I(L1) versus time, and then click on the label of the output node to obtain a plot of V(l) versus time.

Step 5. Verify that the simulation results are correct.

First, comparing the plot in Figure 4.12a to the plot in Figure 4.10. shows that the voltage source represents $v_s(t)$ accurately.

Next, the voltage of the voltage source is equal to the voltage across the capacitor so the capacitor current, $i_c(t)$, is related to $v_s(t)$ by the equation

$$v_L(t) = L\frac{\mathrm{d}}{\mathrm{d}t}i_s(t) = 5\times10^{-3}\frac{\mathrm{d}}{\mathrm{d}t}i_s(t) \qquad (4.10)$$

At any particular time, the value of the derivative of the inductor current is equal to the value of a slope in Figure 4.12a and the value of the capacitor current is equal to a constant value in Figure 4.12b. We readily verify that the inductor voltage represented by Figure 4.12b and the inductor current represented by Figure 4.12a satisfy Eq. (4.10).

Step 6. Report the answer to the circuit analysis problem.

Figure 4.12b is a plot of the inductor voltage, $v_L(t)$, for the circuit in Figure 4.12a.

Example 4.3   (Continued)

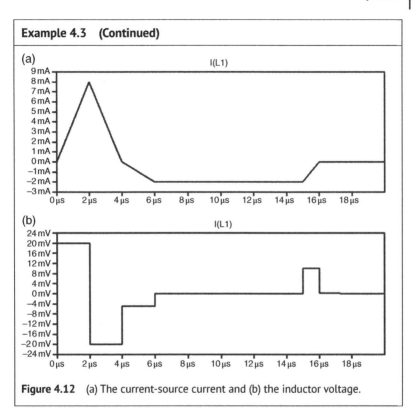

Figure 4.12   (a) The current-source current and (b) the inductor voltage.

References

1 Svoboda, J.A. and Dorf, R.C. (2014). *Introduction to Electric Circuits*, 9e, 8. NJ: Wiley.

2 Halliday, D., Resnick, R., and Walker, J. (2001). *Fundamentals of Physics*, 6e, 515–516. NJ: Wiley.

3 Svoboda, J.A. and Dorf, R.C. (2014). *Introduction to Electric Circuits*, 9e, 280. NJ: Wiley.

Chapter 5

First-Order Circuits

5.1 First-Order Circuits

Circuits that contain capacitors and inductors can be represented by differential equations. The order of the differential equation depends on the number of capacitors and inductors. Circuits that contain only one inductor and no capacitors or only one capacitor and no inductors can be represented by a first-order differential equation. These circuits are called "first-order circuits" [1].

Example 5.1

Consider the first-order circuit shown in Figure 5.1. The input to this circuit is the voltage-source voltage, V_s. Figure 5.1 indicates that $V_s = -15\,mV$ when $t < 0$. Let us assume that the circuit in Figure 5.1 is a DC circuit before the value of V_s changes abruptly to 45 mV. We will see that the abrupt change in the value of V_s will disturb voltages and currents in the circuit. Consequently, the circuit will not be a DC circuit immediately after time $t = 0$. The input voltage will be 45 mV after time $t = 0$, but other voltages and currents will not have constant values. Eventually, the disturbance will die out and all the voltages and currents will again have constant values, likely different constant values than they had before the disturbance. The circuit will again be a DC circuit.

We say that the circuit is "at steady state" before $t = 0$ and again after the disturbance has died out. A "DC circuit" is a circuit in which all the inputs have constant values and the circuit is at steady state. Figure 5.2 summarizes the LTspice analysis of the DC circuits (i) before and (ii) after

(Continued)

LTspice® for Linear Circuits, First Edition. James A. Svoboda.
© 2023 John Wiley & Sons, Inc. Published 2023 by John Wiley & Sons, Inc.

Example 5.1 (Continued)

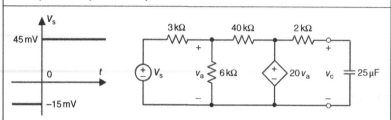

Figure 5.1 The first-order circuit considered in Example 5.1 including a plot of the input voltage, V_s.

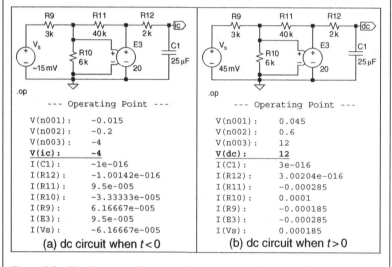

Figure 5.2 Steady-state voltages and currents (a) before and (b) after $t = 0$.

$t = 0$. Each column of Figure 5.2 shows the appropriate LTspice schematic (notice the values of the voltage-source voltages) and the simulation results obtained by performing an "DC op pnt" simulation. In particular, we see from Figure 5.2 that the steady-state value of the capacitor voltage is −4V before $t = 0$ and 12V after $t = 0$.

Step 1. Formulate a circuit analysis problem.

The value of the capacitor voltage, v_c, of the circuit shown in Figure 5.1 makes a transition from the steady-state value of −4V before $t = 0$ to the steady-state value of 12V after $t = 0$. Express the value of the capacitor voltage, v_c, during that transition as a function of time.

Example 5.1 (Continued)

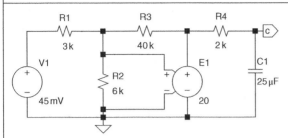

Figure 5.3   The circuit from Figure 5.1 represented as an LTspice schematic.

Step 2. Describe the circuit using LTspice.

Click on the New Schematic toolbar icon, ⬚, to start a new schematic. Place the component symbols and adjust the component values. Place a ground symbol and wire the circuit. Label the output node to obtain the schematic shown in Figure 5.3.

(The value of the input voltage is 45 mV in this schematic of Figure 5.3 because the value of the input voltage will be 45 mV while the capacitor voltage makes the transition from −4 to 12 V.)

Step 3. Simulate the circuit using LTspice.

Click on the Edit Tab, and then select "Spice Analysis" to pop up the Edit Simulation Command dialog box to pop up the "New Simulation" dialog box. Specify a "Transient" simulation, and then specify a "Stop Time" of 250 ms to request a simulation that starts at time 0 and ends at time 250 ms. Click the "OK" button to close the "New Simulation" dialog box. LTspice will generate a "spice directive" consisting of the command ".tran 250 ms." Position the spice directive on the schematic as desired.

The transient response of a circuit is a response to both the input to the circuit and to the "initial conditions" of capacitors and/or inductors in the circuit. The initial condition of a capacitor is the value of the capacitor voltage at time $t = 0$, and the initial condition of an inductor is the value of the inductor current at time $t = 0$. In the present case, the initial capacitor voltage is $v_c(0) = -4$ V. We include the initial condition in the LTspice schematic in one of the following two ways:

1) Click on the SPICE directive icon, $^{.op}$, at the right end of the LTspice toolbar to open a dialog box. Enter

   $$.ic\ V(c) = -4$$

   to generate a "spice directive." Position the spice directive on the schematic as desired.

(Continued)

Example 5.1 (Continued)

2) <CNTL><Right>click on the capacitor. Type "IC = −4" in the "Value" field of the "SpiceLine" Attribute. Optionally, type "X" in the "Vis." field to make the initial condition visible on the schematic.

Click on the Run LTspice toolbar icon, 🏃, to simulate the circuit.

Step 4. Display the results of the simulation.

An empty plot displays automatically. Click on the label of the output node of the schematic to display the plot of the capacitor voltage shown in Figure 5.4.

The simulation results in Figure 5.4 show that $v_c(t)$ makes an exponential transition from −4 V, the steady-state capacitor voltage before $t = 0$, to 12 V, the steady-state capacitor voltage after $t = 0$. Consequently, $v_c(t)$ is represented by an equation of the form

$$v_c(t) = A + Be^{-at}$$

where the values of the parameters A, B, and a are determined from the simulation results. In particular

$$-4 = v_c(0) = A + Be^{-a(0)} = A + B$$

$$12 = \lim_{t \to \infty} A + Be^{-a(0)} = A$$

and

$$a = \frac{\ln\left(\dfrac{A - v_c(t)}{B}\right)}{-t}$$

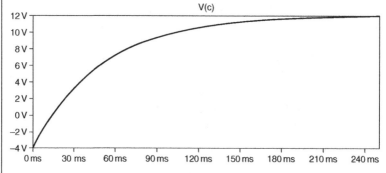

V(c)

Figure 5.4 LTspice transient response of the circuit shown in Figure 5.1.

Example 5.1 (Continued)

First, we get $A = 12$ and $B = -16$.

Next, click on the label "V(c)" at the top of the plot to obtain a cursor. Position the cursor at $t = 60$ ms and observe that $v_c(0.06) = 7.18$ V. Consequently,

$$a = \frac{\ln\left(\dfrac{12-7.18}{16}\right)}{-0.06} = 20$$

and

$$v_c(t) = 12 - 16e^{-20t} \cdot V$$

Figure 5.5 shows plots of $v_c(t)$ given by this equation and $v_c(t)$ obtained by LTspice simulation on the same set of axis. The two plots are identical, one lying on top of the other, confirming that the equation does indeed represent the simulation results.

(We add the equation representing $v_c(t)$ to the simulation results as follows:

1) Right-click on the plot and choose "Add Trace" from the pop-up list.
2) Select "time" from the list of "Available Data."
3) Edit the "Expression to add:" to obtain "12 − 16*exp(−20*time)."
4) Hit <Return>.)

Step 5. Verify that the simulation results are correct.

The state variable in a first-order circuit containing a capacitor is the voltage across that capacitor. Similarly, the state variable in a first-order circuit containing an inductor is the current in that inductor.

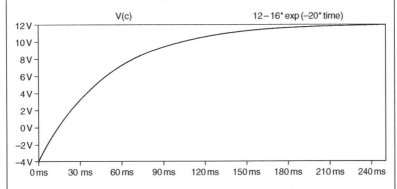

Figure 5.5   Plot of the equation representing $v_c(t)$ superimposed on the plot of $v_c(t)$ itself.

(Continued)

Example 5.1 (Continued)

Table 5.1 Summary of first-order circuits.

First-order circuits containing a capacitor	First-order circuits containing an inductor
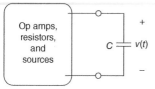	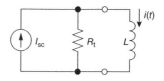
Replace the part of the circuit connected to the capacitor by its Thevenin equivalent circuit:	Replace the part of the circuit connected to the capacitor by its Thevenin equivalent circuit:
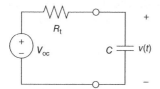	
The capacitor voltage is given by $v(t) = V_{OC} + (v(0) - V_{OC})e^{-at}$	The inductor current is given by $i(t) = I_{SC} + (i(0) - I_{SC})e^{-at}$
where	where
$a = \dfrac{1}{\tau}$ and $\tau = R_t C \Rightarrow a = \dfrac{1}{R_t C}$	$a = \dfrac{1}{\tau}$ and $\tau = \dfrac{L}{R_t} \Rightarrow a = \dfrac{R_t}{L}$

Table 5.1 shows how to obtain an equation representing the voltage across the capacitor in a first-order circuit after determining the Thevenin equivalent of the part to the circuit connected to the capacitor.

Figure 5.6 displays the LTspice schematics and simulation results involved in determining the values of the open circuit voltage, V_{oc}, and the short circuit current, I_{sc}, of that Thevenin equivalent circuit. We see from Figure 5.6 that

$$V_{oc} = 12\,V \quad \text{and} \quad I_{sc} = 6\,mA$$

Example 5.1 (Continued)

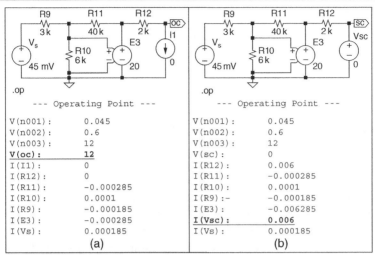

(a)

(b)

Figure 5.6 LTspice simulation results. (a) Voc. (b) Isc.

Consequently, the Thevenin resistance of the part of the circuit connected to the capacitor is

$$R_t = V_{oc}/I_{sc} = 2\,k\Omega$$

From Table 5.1 we see that $\tau = R_t\,C = (2\,k\Omega)*(25\,\mu F) = 50\,ms$ and

$$a = \frac{1}{\tau} = \frac{1}{50\times10^{-3}} = 20\frac{1}{s}$$

and, finally

$$v_c(t) = 12 - 16e^{-20t}$$

This is the same equation obtained earlier and plotted in Figure 5.5, so we are confident that it accurately represents the voltage across the capacitor in the circuit shown in Figure 5.1.

Step 6. Report the answer to the circuit analysis problem.

The voltage across the capacitor in the circuit shown in Figure 5.1 is given by

$$v_c(t) = 12 - 16e^{-20t}\ V$$

Example 5.2

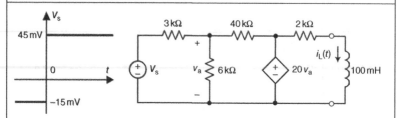

Figure 5.7 The first-order circuit considered in Example 5.2 including a plot of the input voltage, V_s.

Next we consider the circuit shown in Figure 5.7. This circuit is quite similar to the circuit shown in Figure 5.1 and analyzed in Example 5.1. The 25 µF capacitor in Figure 5.1 has been replaced by a 100 mH inductor in Figure 5.7 but otherwise the two circuits are the same. In particular, the part of the circuit connected to the inductor in Figure 5.7 is identical to the part of the circuit connected to the capacitor in Figure 5.1. We have already determined the parameters of the Thevenin equivalent of this part of the circuit in Example 5.1. Recall that

$$V_{oc} = 12\,V, I_{sc} = 6\,mA, \quad \text{and} \quad R_t = V_{oc}/I_{sc} = 2\,k\Omega$$

Referring to the right column of Table 5.1, we see that the current in the inductor in Figure 5.7 is given by

$$i(t) = I_{sc} + \left(i(0) - I_{sc}\right)e^{-at}$$

where

$$a = \frac{R_t}{L} = \frac{2\,k\Omega}{100\,mH} = 20000\,\frac{1}{s}$$

so

$$i_L(t) = 6 + \left(i(0) - 6\right)e^{-20000t}\ mA \tag{5.1}$$

Step 1. Formulate a circuit analysis problem.

Express the value of the inductor current, $i_L(t)$, of the circuit shown in Figure 5.7 as a function of time for $t > 0$.

Example 5.2 (Continued)

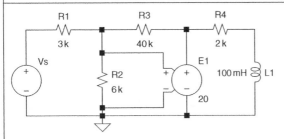

Figure 5.8 The circuit from Figure 5.7 represented as an LTspice schematic.

Step 2. Describe the circuit using LTspice.

Click on the New Schematic toolbar icon, ⌖, to start a new schematic. Place the component symbols, adjust the component values, place a ground symbol, and wire the circuit to obtain the schematic shown in Figure 5.8.

Let us assume that the circuit in Figure 5.7 is a DC circuit before time $t = 0$. Set $V_s = -15$ mV, the value of the voltage-source voltage when $t < 0$, and perform a "DC op pnt" simulation.

The results of the "DC op pnt" simulation are:

```
V(n001):      -0.015                voltage
V(n002):      -0.2                  voltage
V(n003):      -4                    voltage
V(n004):      -2e-006               voltage
I(L1):        -0.002                device_current
I(R4):        -0.002                device_current
I(R3):        9.5e-005              device_current
I(R2):        -3.33333e-005         device_current
I(R1):        6.16667e-005          device_current
I(E1):        0.002095              device_current
I(Vs):        -6.16667e-005         device_current
```

(A little work is required to convince ourselves that I(L1) = −2 mA is the current **directed downward** in the inductor. Notice that n001, n002, n003, and n004 are the names of the four nodes, from left to right, across the top of the circuit. Apply Ohm's law to resistor R4 to see that −2 mA is the current directed from left to right in R4. Apply KCL at the top node of the inductor to see that I(L1) = −2 mA is the current directed downward in the inductor.)

(Continued)

Example 5.2 (Continued)

The initial current in the inductor is

$$i(0) = I(L1) = -2\,\text{mA} \tag{5.2}$$

Consequently

$$
\begin{aligned}
i_L(t) &= 6 + \left(i(0) - 6\right)e^{-20000t} = 6 + (-2-6)e^{-20000t} \\
&= 6 - 8e^{-20000t}\,\text{mA}
\end{aligned} \tag{5.3}
$$

Click on the SPICE directive icon, ·op, at the right end of the LTspice toolbar to open a dialog box. Enter

$$.\text{ic } I(L1) = -2\,\text{mA}$$

to generate a "spice directive." Position the spice directive on the schematic as desired.

Change input voltage to 45 mV and specify a "Transient" simulation with a "Stop Time" to 250 μs to obtain the LTspice schematic shown in Figure 5.9.

Step 3. Simulate the circuit using LTspice.

Click on the Run LTspice toolbar icon, 🏃, to simulate the circuit.

Step 4. Display the results of the simulation.

An empty plot displays automatically. Click on the label of the output node of the schematic to display the plot of the inductor current shown in Figure 5.10.

Step 5. Verify that the simulation results are correct.

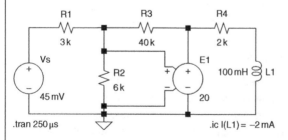

Figure 5.9 LTspice schematic for the transient response of the circuit in Figure 5.8.

Example 5.2 (Continued)

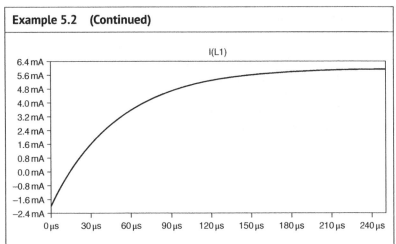

Figure 5.10 LTspice transient response of the circuit shown in Figure 5.8.

Equation (5.3) indicates that

$$i_L(t) = 6 - 8e^{-20000t} \text{ mA} \tag{5.4}$$

We can add a plot of this expression for $i_L(t)$ to Figure 5.10 as follows:

1) Right-click the plot and choose "Add Trace" from the pop-up list.
2) Select "time" from the list of "Available Data."
3) Edit the "Expression to add:" to obtain "6 – 8*exp(–20000*time)."
4) Hit <Return>.
   Click on the Run LTspice toolbar icon, 🏃, to simulate the circuit and display the plot shown in Figure 5.11.

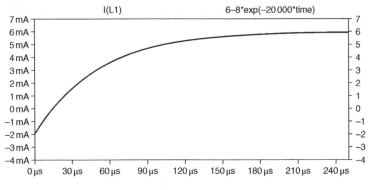

Figure 5.11 Plots of LTspice simulations of both sides of Eq. (5.2).

(Continued)

Example 5.2 (Continued)

Figure 5.11 shows plots of both $i_L(t)$ obtained by LTspice simulation and $i_L(t)$ given by Eq. (5.3) on the same set of axis. The two plots are identical, one lying on top of the other, confirming that the equation does indeed represent the simulation results.

Step 6. Report the answer to the circuit analysis problem.

The current in the inductor in the circuit shown in Figure 5.10 is given by

$$i_L(t) = 6 - 8e^{-20000t} \text{ mA}$$

Example 5.3

Step 1. Formulate a circuit analysis problem.

Assume that the circuit shown in Figure 5.12 is a DC circuit before $t = 0$. Express the value of the capacitor voltage, v_c, as a function of time for $t > 0$.

Step 2. Describe the circuit using LTspice.

Click on the New Schematic toolbar icon, ⌁, to start a new schematic. Place the component symbols and adjust the component values. Place a ground symbol. Wire the circuit. Label the output node to obtain the schematic shown in Figure 5.13. This schematic represents the DC circuit before $t = 0$. (Notice that the value of the input voltage is 0V.) Simulate this DC circuit to determine that the capacitor voltage is 0V at time $t = 0$.

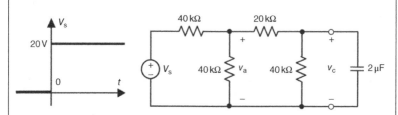

Figure 5.12 The first-order circuit considered in Example 5.1 including a plot of the input voltage, V_s.

Example 5.3 (Continued)

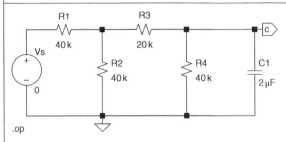

Figure 5.13 The circuit from Figure 5.12 *before t* = 0 represented as an LTspice schematic.

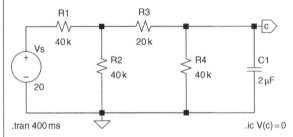

Figure 5.14 The circuit from Figure 5.12 *after t* = 0 represented as an LTspice schematic.

The schematic in Figure 5.14 represents the circuit after time *t* = 0 when the circuit is not a DC circuit. Notice that a transient response is specified and the initial capacitor voltage is specified.

Step 3. Simulate the circuit using LTspice.

Click on the Run LTspice toolbar icon, 🏃, to simulate the circuit.

Step 4. Display the results of the simulation.

An empty plot displays automatically. Click on the label of the output node of the schematic to display the plot of the inductor current shown in Figure 5.15.

Step 5. Verify that the simulation results are correct.

We can use Table 5.1 to determine an equation representing the capacitor voltage in Figure 5.15. To do so, we need to determine the Thevenin equivalent circuit of the part of the circuit that is connected to the capacitor.

Figure 5.16 displays the LTspice schematics and simulation results involved in determining the values of the open circuit voltage, V_{oc}, and the short circuit current, I_{sc}, of the Thevenin equivalent circuit of the

(Continued)

Example 5.3 (Continued)

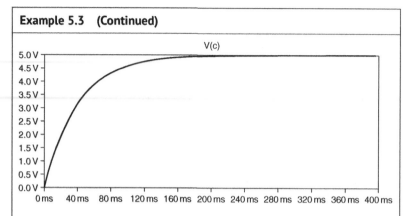

Figure 5.15 LTspice transient response of the circuit shown in Figure 5.12.

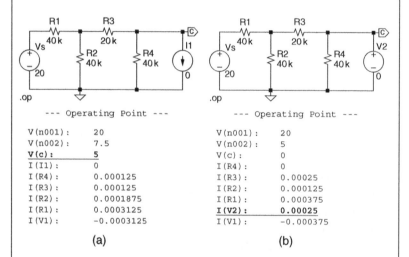

Figure 5.16 Thevenin equivalent of the part of the circuit connected to the capacitor. (a) Voc. (b) Isc.

part of the circuit in Figure 5.14 that is connected to the capacitor. We see from Figure 5.16 that

$$V_{oc} = 5\,V \quad \text{and} \quad I_{sc} = 0.25\,mA$$

Consequently, the Thevenin resistance of the part to the circuit connected to the capacitor is

$$R_t = V_{oc}/I_{sc} = 20\,k\Omega$$

Example 5.3 (Continued)

From Table 5.1 we see that $\tau = R_t\,C = (20\,k\Omega)^*(2\,\mu F) = 40\,ms$ and

$$a = \frac{1}{\tau} = \frac{1}{40 \times 10^{-3}} = 25\frac{1}{s}$$

and, finally

$$v_c(t) = 5 - 5e^{-25t} \tag{5.5}$$

We can add a plot of this expression for $i_L(t)$ to Figure 5.15 as follows:

1) Right-click the plot and choose "Add Trace" from the pop-up list.
2) Select "time" from the list of "Available Data."
3) Edit the "Expression to add:" to obtain "5 – 5*exp(–25*time)."
4) Hit <Return>.

Click on the Run LTspice toolbar icon, ✗, to simulate the circuit and display the plot shown in Figure 5.17.

Figure 5.17 shows plots of both $v_C(t)$ obtained by LTspice simulation and $i_L(t)$ given by Eq. (5.5) on the same set of axis. The two plots are identical, one lying on top of the other, confirming that the equation does indeed represent the simulation results.

Step 6. Report the answer to the circuit analysis problem.

The voltage across the capacitor in the circuit shown in Figure 4.10 is given by

$$v_c(t) = 5 - 5e^{-25t} \text{ V}$$

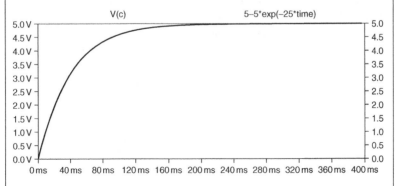

Figure 5.17 Plots of LTspice simulations of both sides of Eq. (5.5).

Example 5.4

Consider again the circuit shown in Figure 5.12 and reproduced for convenience in Figure 5.18. The resistor voltage, v_a, is the output of this circuit. As such, it can be expressed as a linear combination of the input, V_s, and the state variable, v_c.

$$v_a = a_1 V_s + a_2 v_c \tag{5.6}$$

From the plot in Figure 5.18, we see that $V_s = 20\text{V}$ after time $t = 0$. In Example 5.3 we saw that $v_c = 5(1-e^{-25t})\text{V}$ after time $t = 0$. Substituting these expressions into Eq. (5.6) gives

$$v_a = a_1(20) + a_2\left(5\left(1-e^{-25t}\right)\right) = a_1(20) + a_2\left(5 - 5e^{-25t}\right) \tag{5.7}$$

Step 1. Formulate a circuit analysis problem.

Assume that the circuit shown in Figure 5.18 is a DC circuit before $t = 0$. Express the value of the output voltage, v_a, as a function of time for $t > 0$.

Step 2. Describe the circuit using LTspice.

Click on the New Schematic toolbar icon, 🖸, to start a new schematic. Place the component symbols and adjust the component values. Place a ground symbol. Wire the circuit. Label the output node as node "a" and the capacitor node as node "c" to obtain the schematic shown in Figure 5.19. This schematic represents the circuit after $t = 0$. (Notice that the value of the input voltage is 20V because the schematic represents the circuit when $t > 0$. Also notice that value of the initial capacitor voltage is 0V because the values of all the voltages are 0V before time $t = 0$.)

Step 3. Simulate the circuit using LTspice.

Click on the Run LTspice toolbar icon, 🏃, to simulate the circuit.

Step 4. Display the results of the simulation.

An empty plot displays automatically. Click on the labels of both nodes "a" and "b" of the schematic to display the plot shown in Figure 5.20.

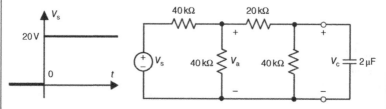

Figure 5.18 The first-order circuit considered in Example 5.3 including a plot of the input voltage, V_s.

Example 5.4 (Continued)

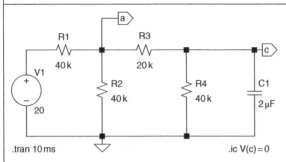

Figure 5.19 The circuit from Figure 5.18 *after t = 0* represented as an LTspice schematic.

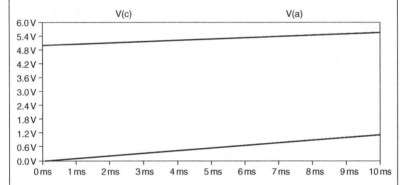

Figure 5.20 The circuit from Figure 5.19 *after t = 0* represented as an LTspice schematic.

Use the cursor provided by LTspice to measure the values of v_a and v_c at times $t = 0$ and $t = 7$ ms:

$$v_a = 5V \quad \text{and} \quad v_c = 0V \quad \text{at} \quad t = 0 \tag{5.8}$$

$$v_a = 5.4V \quad \text{and} \quad v_c = 0.8V \quad \text{at} \quad t = 7\text{ms} \tag{5.9}$$

Substituting the values of v_a and v_c from Eq. (5.8) into Eq. (5.7) gives

$$5 = a_1(20) + a_2(0) \implies a_1 = \frac{1}{4}$$

Next, substituting the values of v_a and v_c from Eq. (5.9) into Eq. (5.7) gives

$$5.4 = \frac{1}{4}(20) + a_2(0.8) \implies a_2 = \frac{5.4 - 5}{0.8} = \frac{1}{4}$$

(Continued)

Example 5.4   (Continued)

Consequently

$$v_a = \frac{1}{4}(20) + \frac{1}{2}\left(5 - 5e^{-25t}\right) = 7.5 - 5e^{-25t} \tag{5.10}$$

Step 5. Verify that the simulation results are correct.

Figure 5.19 shows a schematic that we can use to plot $v_a(t)$ as a function of time for $t > 0$. The stop time of the transient response is 10 ms in this schematic. Change the stop time of the transient response to 400 ms. Click on the Run LTspice toolbar icon, 🏃, to simulate the circuit and display the plot.

Figure 5.21 shows plots of both $v_a(t)$ obtained by LTspice simulation and $v_a(t)$ given by Eq. (5.10) on the same set of axis. The two plots are identical, one lying on top of the other, confirming that the equation does indeed represent the simulation results.

Step 6. Report the answer to the circuit analysis problem.

The resistor voltage, v_a, in the circuit shown in Figure 5.18 is given by

$$v_a(t) = 7.5 - 2.5e^{-25t} \text{ V}$$

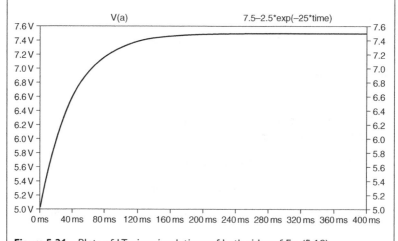

Figure 5.21   Plots of LTspice simulations of both sides of Eq. (5.10).

Reference

1 Svoboda, J.A. and Dorf, R.C. (2014). *Introduction to Electric Circuits*, 9e, 322. NJ: Wiley.

Chapter 6

AC Circuits

An "AC circuit" is a "steady-state" linear circuit that has sinusoidal inputs, all having the same frequency. The sinusoids can be represented either as sines or as cosines. These sinusoids will have frequencies with units of either hertz or radians/second. LTspice, and SPICE in general, represents sinusoids as sines with frequencies in hertz. We will have no trouble making the necessary conversions, but we do need to remember to make them.

6.1 Phasors

Figure 6.1 shows an AC circuit having sinusoidal inputs represented as cosines. Both inputs have the same frequency:

1000π rad/s = 500 Hz

Apply KCL at the top node of the resistor, and then use Ohm's law to obtain

$$v(t) = 10\left(5\cos\left(1000\pi t + 30°\right) + 4\cos\left(1000\pi t + 60°\right)\right) \tag{6.1}$$

We can use the trigonometric identity

$$\cos\left(\alpha + \beta\right) = \cos\left(a\right)\cos\left(\beta\right) - \sin\left(a\right)\sin\left(\beta\right)$$

LTspice® for Linear Circuits, First Edition. James A. Svoboda.
© 2023 John Wiley & Sons, Inc. Published 2023 by John Wiley & Sons, Inc.

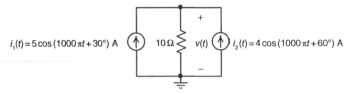

Figure 6.1 An AC circuit.

to solve Eq. (6.1) as follows:

$$v(t) = 10\big(5\cos(1000\pi t + 30°) + 4\cos(1000\pi t + 60°)\big)$$
$$= 50\big(\cos(1000\pi t)\cos(30°) - \sin(1000\pi t)\sin(30°)\big)$$
$$+ 40\big(\cos(1000\pi t)\cos(60°) - \sin(1000\pi t)\sin(60°)\big) \quad (6.2a)$$
$$= \cos(1000\pi t)\big[50\cos(30°) + 40\cos(60°)\big]$$
$$- \sin(1000\pi t)\big[50\sin(30°) + 40\sin(60°)\big]$$
$$= 63.301\cos(1000\pi t) - 59.641\sin(1000\pi t)$$

Now we need another trigonometric identity:

$$A\cos(\alpha) + B\sin(\alpha) = \sqrt{A^2 + B^2}\,\cos\!\left(\alpha + \text{atan}\!\left(\frac{-B}{A}\right)\right)$$

Finally

$$v(t) = \sqrt{63.301^2 + 59.641^2}\,\cos\!\left(1000\pi t + \text{atan}\!\left(\frac{-(-59.641)}{63.301}\right)\right) \quad (6.2b)$$
$$= 86.972\cos(1000\pi t + 43.3°)\,\text{V}$$

That was a lot of work. Usually, we will find it easier to solve equations like Eq. (6.1) using phasors. A **phasor** is a complex number that we associate with a sinusoid [1]:

$$A\angle\theta \iff A\cos(\omega t + \theta)$$

For example, the phasors corresponding to $i_1(t)$ and $i_2(t)$ are

$$5\angle 30° \iff 5\cos(\omega t + 30°) \text{ and } 4\angle 60° \iff 4\cos(\omega t + 60°)$$

Let $V\angle\theta$ denote the phasor corresponding to $v(t)$. Replacing sinusoids in Eq. (6.1) by the corresponding phasors we get

$$V\angle\theta = 10\big(5\angle 30° + 4\angle 60°\big) \quad (6.3)$$

Doing the complex arithmetic [2] on the right-hand side, we get

$$V \angle \theta = 10\left(5 \angle 30° + 4 \angle 60°\right)$$
$$= 10\left(\left(4.3301 + j2.5000\right) + \left(2.0000 + j3.464\right)\right)$$
$$= 10\left(6.3301 + j5.9641\right)$$
$$= 10\left(\sqrt{6.3301^2 + 5.9641^2} \angle \tan^{-1}\left(\frac{5.9641}{6.3301}\right)\right)$$
$$= 86.972 \angle 43.3° \text{ V}$$

(6.4)

As should be expected, the phasor $V \angle \theta$ calculated in Eq. (6.4) is the phasor corresponding to the sinusoid $v(t)$ calculated in Eq. (6.2b).

Equations (6.2a), (6.2b), and (6.3) suggest that phasors corresponding to currents of a circuit obey the KCL equations of that circuit and indeed they do [3]. Similarly, phasors corresponding to voltages of a circuit obey the KVL equations of that circuit.

Figure 6.2 shows how the amplitude, A, and period, T, of a sinusoid can be measured from a plot of the sinusoid. The sinusoid is then represented as shown in Eq. (6.5).

$$v(t) = A\cos(\omega t + \theta) = A\cos\left(\frac{2\pi}{T}t + \theta\right)$$

(6.5)

Next, the value of the phase angle θ can be determined from the coordinates, t_1 and $v(t_1)$, of a point on the sinusoid using the following equation:

$$\theta = \begin{cases} -\cos^{-1}\left(\frac{v(t_1)}{A}\right) - \omega t_1 & \begin{array}{l} \text{when } v(t_1) > 0 \text{ and } v(t) \text{ is increasing at time } t_1 \\ \text{or} \\ \text{when } v(t_1) < 0 \text{ and } v(t) \text{ is decreasing at time } t_1 \end{array} \\ \\ \cos^{-1}\left(\frac{v(t_1)}{A}\right) - \omega t_1 & \begin{array}{l} \text{when } v(t_1) > 0 \text{ and } v(t) \text{ is decreasing at time } t_1 \\ \text{or} \\ \text{when } v(t_1) < 0 \text{ and } v(t) \text{ is increasing at time } t_1 \end{array} \end{cases}$$

(6.6)

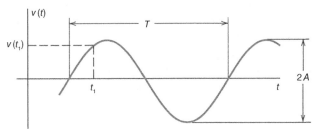

Figure 6.2 A sinusoidal voltage.

In Example 6.1 we will simulate circuit shown in Figure 6.1 to obtain a plot of the sinusoidal voltage $v(t)$. We will measure A, T, t_1, and $v(t_1)$ as shown in Figure 6.2 and calculate θ using Eq. (6.6). We will then have determined $v(t)$ in three different ways: using trigonometry, using phasors, and using LTspice.

Example 6.1

Determine $V\angle\theta$, the phasor of the voltage across the 10 Ω resistor of the circuit shown in Figure 6.1.

Step 1. Formulate a circuit analysis problem.
Plot the voltage $v(t)$ across the 10 Ω resistor of the circuit shown in Figure 6.1.

Step 2. Describe the circuit using LTspice.
LTspice represents sinusoids as sines rather than cosines so, using a trigonometric identity,

$$\cos(\alpha) = \sin(\alpha + 90°)$$

We will represent the current-source currents of the circuit in Figure 6.1 as

$$i_1(t) = 5\sin(1000\pi t + 120°) \quad \text{and} \quad i_2(t) = 4\sin(1000\pi t + 150°)$$

Click on the New Schematic toolbar icon, ⟨⁊⟩, to start a new schematic. Place symbols representing the current sources, resistor, and ground node. Specify the resistance.

Right-click the left current source and click the "Advanced" button in the "Independent Current Source – I1" dialog box, and then specify the current-source waveform as shown in Figure 6.3.

Right-click the other current source and click the "Advanced" button in the "Independent Current Source – I2" dialog box; then specify the current-source current to finish the schematic shown in Figure 6.4.

Step 3. Simulate the circuit using LTspice.
Select the Edit Tab, and then select "Spice Analysis" to pop up the Edit Simulation Command dialog box. Provide a Name and then select Create. The Edit Simulation Command dialog box will pop up. Select the "Transient" tab and specify a "Stop Time" equal to 20 ms and a "Maximum Timestep" equal to 0.01 ms. Select "OK" to close Edit Simulation Command dialog box. Click on the Run LTspice toolbar icon, ⟨⁊⟩, to simulate the circuit.

Example 6.1 (Continued)

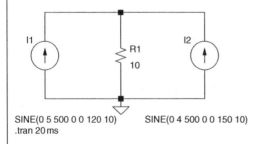

Independent Current Source - I1
Functions

○ (none)

○ PULSE(I1 I2 Tdelay Trise Tfall Ton Period Ncycles)

◉ SINE(Ioffset Iamp Freq Td Theta Phi Ncycles)

○ EXP(I1 I2 Td1 Tau1 Td2 Tau2)

○ SFFM(Ioff Iamp Fcar MDI Fsig)

○ PWL(t1 i1 t2 i2...)

○ PWL FILE: [   ] [Browse]

○ TABLE(v1 i1 v2 i2...)

DC offset[A]:	0
Amplitude[A]:	5
Freq[Hz]:	500
Tdelay[s]:	0
Theta[1/s]:	0
Phi[deg]:	120
Ncycles:	10

[Additional PWL Points]

Make this information visible on schematic: ☑

Figure 6.3 Specify a sinusoidal current source.

I1   R1
     10   I2

SINE(0 5 500 0 0 120 10)   SINE(0 4 500 0 0 150 10)
.tran 20 ms

Figure 6.4 The LTspice schematic for the circuit in Figure 6.1.

Step 4. Display the results of the simulation.

The results of a Transient simulation display automatically as a blank plot. Click on the top node of resistor R1 to obtain a plot of V(n001) versus time shown in Figure 6.5.

(Continued)

Example 6.1 (Continued)

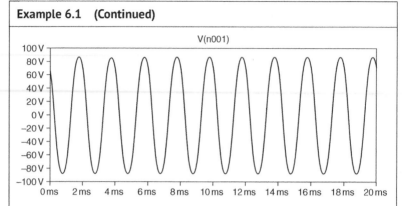

Figure 6.5 Transient response of the simulation in Figure 6.4.

Use the cursor to measure the amplitude, A, and the period, T, of the sinusoid shown in Figure 6.5. (See Figure 6.2 to recall how the period T and amplitude A are defined.) We find

$$A = 86.78\,\text{V and } T = 2.002\,\text{ms}$$

Next we measure the coordinates of a point on the sinusoid in order to calculate the phase angle, θ, using Eq. (6.6). For example

$$t_1 = 15.5094\,\text{ms and } v(t_1) = 61.387638\,\text{V} \Rightarrow \theta = 43.3°$$

Consequently, the steady-state response of the AC circuit in Figure 6.1 is

$$v(t) = 86.78\cos(1000\pi t + 43.3°)\,\text{V}$$

Step 5. Verify that the simulation results are correct.

The simulation results agree nicely with the sinusoid obtained earlier in Eq. (6.2b) and the phasor obtained in Eq. (6.4), so we are confident that the simulation results are correct.

Step 6. Report the result.

The phasor of the voltage across the 10 Ω resistor of the circuit shown in Figure 6.1 is

$$V\angle\theta = 86.78\angle 43.3°\,\text{V}.$$

6.2   AC Circuits

In Example 6.2 we will encounter the circuit in Figure 6.6 in which the input is a sinusoidal voltage but, **initially**, the current and the voltages of the resistor, inductor, and capacitor are not sinusoidal. Consequently, the circuit is not "at steady state." The resistor, inductor, and capacitor current and voltages each contain a nonsinusoidal component. Eventually, these nonsinusoidal components die out, after which the circuit is said to be "at steady state." The circuit is not called an "AC circuit" until it is at steady state.

We expect that eventually this circuit will be at steady state and the current $i(t)$ will be a sinusoidal. Then

$$i(t) = A\cos(\omega t + \theta) = A\cos\left(\frac{2\pi}{T}t + \theta\right) \tag{6.7}$$

The values of the amplitude A and period T can be determined from a plot of $i(t)$ versus t as shown in Figure 6.7. The value of the phase angle θ can be determined from A, T, and the coordinates, t_1 and $i(t_1)$, of a point

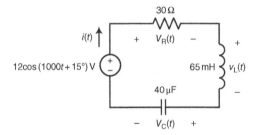

Figure 6.6   An AC circuit.

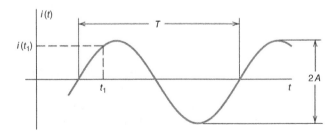

Figure 6.7   A sinusoidal current.

on the sinusoid. After obtaining values of t_1 and $i(t_1)$ as shown in Figure 6.7, the value of the phase angle θ can be determined using the following equation:

$$\theta = \begin{cases} -\cos^{-1}\left(\dfrac{v_2(t_1)}{A}\right) - \omega t_1 & \begin{array}{l}\text{when } i(t_1) > 0 \text{ and } i(t_1) \text{ is increasing at time } t_1 \\ \text{or when } i(t_1) < 0 \text{ and } i(t_1) \text{ is decreasing at time } t_1\end{array} \\[4mm] \cos^{-1}\left(\dfrac{v_2(t_1)}{A}\right) - \omega t_1 & \begin{array}{l}\text{when } i(t_1) > 0 \text{ and } i(t_1) \text{ is decreasing at time } t_1 \\ \text{or when } i(t_1) < 0 \text{ and } i(t_1) \text{ is increasing at time } t_1\end{array} \end{cases}$$

$$(6.8)$$

Example 6.2

Determine the current $i(t)$ in the circuit shown in Figure 6.6.

Step 1. Formulate a circuit analysis problem.

Determine the current $i(t)$ in the circuit shown in Figure 6.6.

Step 2. Describe the circuit using LTspice.

LTspice represents sinusoids as sines rather than cosines so, using a trigonometric identity, we will represent the voltage-source voltage of the circuit in Figure 6.6 as

$$12\cos(1000t + 15°) = 12\sin(1000t + 15° + 90°)$$
$$= 12\sin(1000t + 105°)$$

Click on the New Schematic toolbar icon, 🖉, to start a new schematic. Place symbols representing the voltage source, resistor, inductor, and capacitor. Specify the values of the resistance, capacitance, and inductance. Right-click the voltage source and click the "Advanced" button in the "Voltage Source – V1" dialog box, and then specify the voltage-source waveform as shown in Figure 6.8.

Step 3. Simulate the circuit using LTspice.

Click on the Edit Tab, and then select "Spice Analysis" to pop up the Edit Simulation Command dialog box. Select the "Transient" tab, specify the "Stop Time" to be 80 ms, and then click the "OK" button. LTspice will generate a "spice directive" consisting of the command ".trans 80 ms." Position the spice directive on the schematic as shown in Figure 6.9. Click on the Run LTspice toolbar icon, ⚡, to simulate the circuit.

Example 6.2   (Continued)

⊿ Independent Voltage Source - V1

Functions

○ (none)

○ PULSE(V1 V2 Tdelay Trise Tfall Ton Period Ncycles)

◉ SINE(Voffset Vamp Freq Td Theta Phi Ncycles)

○ EXP(V1 V2 Td1 Tau1 Td2 Tau2)

○ SFFM(Voff Vamp Fcar MDI Fsig)

○ PWL(t1 v1 t2 v2...)

○ PWL FILE: [   ] [Browse]

DC offset[V]: 0

Amplitude[V]: 12

Freq[Hz]: 159

Tdelay[s]: 0

Theta[1/s]: 0

Phi[deg]: 105

Ncycles: 14

[Additional PWL Points]

Make this information visible on schematic: ☑

Figure 6.8   Specify the voltage source.

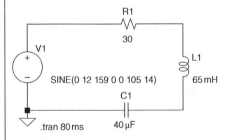

Figure 6.9   The circuit from Figure 6.6 represented as an LTspice schematic.

Step 4. Display the results of the simulation.

The results of a Transient simulation display automatically as a blank plot. Click on the inductor in the schematic to obtain a plot of I(L1) versus time shown in Figure 6.10.

(*Continued*)

Example 6.2   (Continued)

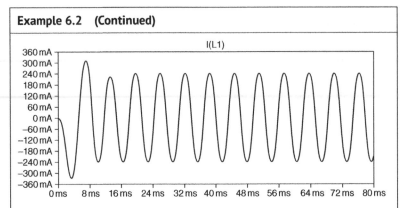

Figure 6.10   The transient response generated by LTspice schematic in Figure 6.9.

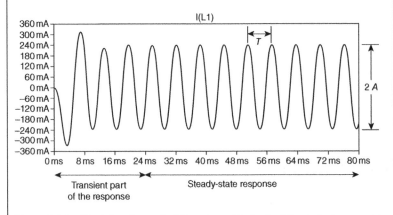

Figure 6.11   Obtaining the period, *T*, and amplitude, *A*, of the inductor current from the transient response.

Notice that the inductor current is not sinusoidal immediately after time $t = 0$. Apparently, the circuit is not at steady state. This plot consists of two parts. One part vanishes as time increases. The part that remains is the "steady-state response." In Figure 6.11 the response is labeled to separate "the transient part of the response," which lasts about by 24 ms, from the "steady-state response."

We can use the cursor to measure the amplitude, *A*, and the period, *T*, of the steady-state response as illustrated in Figure 6.11. We find

$$A = 240\,\text{mA} \text{ and } T = 6.29\,\text{ms}$$

Example 6.2 (Continued)

Next, we measure the time and inductor current corresponding to a point on the steady-state response and notice whether the value of the current is increasing or decreasing at that time. For example

At $t_1 = 62.52$ ms and $i(t_1) = 122$ mA and $i(t)$ is increasing.

Using Eq. (6.6), the phase angle is calculated to be $-37.7°$. Consequently

$$i(t) = 240\cos(1000\pi t - 37.7°)\,\text{mA}$$
$$= 240\sin(1000\pi t + 52.3°)\,\text{mA} \tag{6.9}$$

Similarly, the steady-state voltages across the resistor, capacitor, and inductor can be determined to be

$$v_R(t) = 7.184\cos(1000\pi t - 38.0°)\,\text{V}, v_C(t) = 5.982$$
$$\cos(1000\pi t - 127.6°)\,\text{V, and } v_L(t) = 15.558 \tag{6.10}$$
$$\cos(1000\pi t + 52.3°)\,\text{V}$$

(See the appendix at the end of this chapter for details.)

Step 5. Verify that the simulation results are correct.

In Steps 3 and 4 we used an LTspice **transient analysis** to obtain plots of the current and voltages of the circuit shown in Figure 6.6. Next, we used Figure 6.7 and Eqs. (6.5) and (6.6) to represent these currents and voltages as sinusoidal functions of time in Eqs. (6.9) and (6.10).

In Example 6.3, we will perform an **AC analysis** on the same circuit, using LTspice, to obtain the phasors representing the node voltages:

$$\mathbf{V_a} = 12\angle 105°\,\text{V}, \mathbf{V_b} = 9.6\angle 141.9°\,\text{V, and } \mathbf{V_c} = 6.0\angle -38.1°\,\text{V}$$

The corresponding phasor voltages across the capacitor, resistor, and inductor are

$$\mathbf{V_C} = \mathbf{V_c} = 6.0\angle -38.1°\,\text{V}, \mathbf{V_R} = \mathbf{V_a} - \mathbf{V_b} = 7.2\angle 51.9°\,\text{V, and}$$
$$\mathbf{V_L} = \mathbf{V_b} - \mathbf{V_c} = 15.6\angle 141.9°\,\text{V}$$

Recalling that LTspice associates phasors with sines rather than cosines, we have

$$\mathbf{V_C} = 6\angle -38.1°\,\text{V} \implies v_C(t) = 6\sin(1000\pi t - 38.1°)$$
$$= 6\cos(1000\pi t - 128.1°)\,\text{V}$$

(Continued)

Example 6.2   (Continued)

$$\mathbf{V}_R = 7.2\angle 51.6°\text{V} \quad \Rightarrow \quad v_R(t) = 7.2\sin(1000\pi t + 51.6°)$$
$$= 7.2\cos(1000\pi t - 38.1°)\text{V}$$

$$\mathbf{V}_L = 15.6\angle 141.9°\text{V} \quad \Rightarrow \quad v_L(t) = 15.6\sin(1000\pi t + 141.9°)$$
$$= 7.2\cos(1000\pi t + 51.9°)\text{V}$$

These phasors correspond to the sinusoids in Eq. (6.10), verifying that the sinusoids are correct.

Step 6. Report the answer to the circuit analysis problem.

The current in the circuit shown in Figure 6.6 is

$$i(t) = 240\cos(1000\pi t - 37.7°)\text{mA} = 240\sin(1000\pi t + 52.3°)\text{mA}$$

6.3   Impedances

The **impedance** of a resistor, capacitor, or inductor in an AC circuit is the ratio of its voltage phasor to its current phasor as shown on the right-hand column of Table 6.1. An AC circuit is said to be represented in the **frequency domain** when its currents and voltages are represented as phasors and resistors, capacitors, and inductors are represented by their impedances. In contrast, an AC circuit is said to be represented in the time domain when voltages and currents are represented as sinusoidal functions of time and resistors, capacitor, and inductor are represented as shown on the left-hand column of Table 6.1.

Figure 6.12 shows the circuit from Example 6.2 represented in the time domain (as it was in Figure 6.6) and in the frequency domain. The frequency of this circuit is $500\,\text{Hz} = 1000\pi\,\text{rad/second}$. The impedances of the $40\,\mu\text{F}$ capacitor and the $65\,\text{mH}$ inductor are

$$\mathbf{Z}_C = \frac{1}{j\omega C} = \frac{1}{j(1000\pi)(40\times 10^{-6})} = -j25\Omega$$

and

$$\mathbf{Z}_L = j\omega L = j(1000\pi)(65\times 10^{-3}) = j65\Omega \tag{6.11}$$

In Example 6.3 we analyze the circuit in Figure 6.12 using the LTspice "AC Analysis" simulation. This type of simulation requires a description of an AC circuit, primarily in the time domain, and determines currents and voltages of the circuit represented in the frequency domain. In particular, the user is not required to calculate the impedances of capacitors and inductors, LTspice will.

Table 6.1 Time domain versus frequency representations of AC circuits.

Time domain	Frequency domain
$v(t) = R\,i(t)$, $Z_R = R$	$V(\omega) = Z_R I(\omega)$
$i(t) = C\dfrac{dv(t)}{dt}$	$V(\omega) = Z_C I(\omega)$, $Z_C = \dfrac{1}{j\omega C}$
$v(t) = L\dfrac{di(t)}{dt}$	$V(\omega) = Z_L I(\omega)$, $Z_L = j\omega L$

$Z_R(\omega)$, $Z_C(\omega)$, and $Z_L(\omega)$ are impedances; $i(t)$ and $v(t)$ are sinusoidal functions of time; $I(\omega)$ and $V(\omega)$ are phasors.

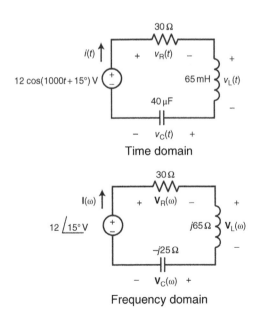

Figure 6.12 The circuit from Example 6.2 represented in the time domain and in the frequency domain. (Note we already saw the time domain circuit in Figure 6.6.)

Example 6.3

Determine the impedances of the capacitor and inductor of the circuit shown in Figure 6.12.

Step 1. Formulate a circuit analysis problem.

Analyze the circuit shown in Figure 6.12a to determine the impedance of the capacitor and inductor.

Step 2. Describe the circuit as an LTspice schematic.

Click on the New Schematic toolbar icon, 🖉, to start a new schematic. Place symbols representing the voltage source, resistor, capacitor, and inductor.

Specify the values of the resistance, capacitance, and inductance. Place a ground symbol and wire the circuit. Label the nodes.

Right-click the voltage source and click the "Advanced" button in the "Voltage Source – V1" dialog box (shown in Figure 6.13), and then specify the amplitude and phase of the voltage-source voltage as shown in Figure 6.15.

Step 3. Simulate the circuit using LTspice.

Select **Simulate/Edit Simulation Cmd** from the LTspice menu bar to pop up the "Edit Simulation Command" dialog box. Specify an "AC Analysis" simulation, and then specify the frequency as a list consisting

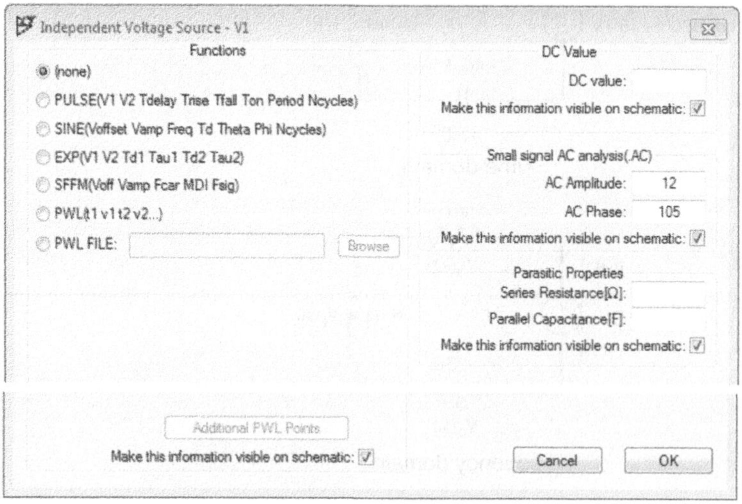

Figure 6.13 Independent voltage source dialog box.

Example 6.3   (Continued)

of the single frequency, 159.155 Hz = 1000 rad/second, as shown in Figure 6.14. (Notice that the frequency of the AC circuit is specified in Hertz instead of rad/second.) LTspice will generate a "spice directive" consisting of the command ".ac list 159.155." Position the spice directive on the schematic as shown in Figure 6.15.

Click on the LTspice toolbar Run icon, ⚡, to simulate the circuit.

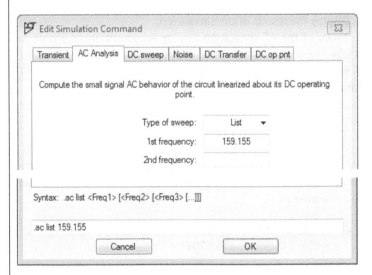

Figure 6.14   AC source frequency dialog box.

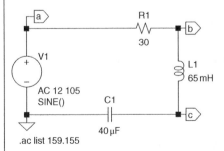

Figure 6.15   LTspice schematic for the AC circuit shown in Figure 6.12.

Example 6.3 (Continued)

```
                    --- AC Analysis ---
frequency:     159.155    Hz
V(c):             mag:  5.99992  phase:   -38.1292°      voltage
V(b):             mag:  9.59989  phase:   141.869°       voltage
V(a):             mag:       12  phase:       105°       voltage
I(C1):            mag: 0.239997  phase:   51.8708°       device_current
I(L1):            mag: 0.239997  phase:   51.8708°       device_current
I(R1):            mag: 0.239997  phase:   51.8708°       device_current
I(V1):            mag: 0.239997  phase:  -128.129°       device_current
```

Figure 6.16 Simulation results.

Step 4. Display the results of the simulation.

The results of the AC simulation display automatically as shown in Figure 6.16.

Step 5. Verify that the simulation results are correct.

The impedance of the capacitor [4] in Figure 6.12 is calculated as

$$Z_c = \frac{1}{j\omega C} = \frac{1}{j(1000\pi)(40 \times 10^{-6})} = -j25\angle-90° = -j25\ \Omega$$

The voltage across the capacitor in Figure 6.12 is equal to the node voltage at node c. Consequently, the impedance of the capacitor can also be calculated as

$$Z_c = \frac{V(c)}{I(C1)} = \frac{5.99992\angle-38.1292°\ V}{0.239997\angle51.8708°\ V} = 24.99998\angle-90°$$
$$= -j24.99998\Omega$$

Since the same value of the capacitor impedance is obtained in both calculations, the simulation results appear to be correct.

The impedance of the inductor in Figure 6.12 is calculated as

$$Z_L = j\omega L = j(1000\pi)(65 \times 10^{-3}) = j65\Omega$$

The impedance of the inductor can also be calculated as

$$Z_L = \frac{V(b) - V(c)}{I(L1)}$$
$$= \frac{(9.59989\angle141.869°)\ V - (5.99992\angle-38.1292°)\ V}{0.239997\angle51.8708°\ V} = -j65\Omega$$

Example 6.3 (Continued)

Since the same value of the inductor impedance is obtained in both calculations, the simulation results appear to be correct.

Step 6. Report the answer to the circuit analysis problem.

The impedance of the capacitor in Figure 6.12 is $-j25\,\Omega$.

The impedance of the inductor in Figure 6.12 is $j65\,\Omega$.

6.A Appendix: Further Details for Example 6.2

Figure 6.A.1 shows the LTspice schematic from Figure 6.9 after labeling the node. Click on the Run LTspice toolbar icon, $\mathcal{K}$, to simulate this circuit. The results of a Transient simulation display automatically as a blank plot. Click on the label of node b to obtain a plot of V(b) versus time. Next, *right-click* the label "V(b)" in the space at the top of the plot. A dialog box will provide an opportunity to change "V(b)" to "V(b) − V(c)." Do so, and then click "OK" to dismiss the dialog box. The plot will change to display V(b) − V(c), which is the inductor voltage.

Use the cursor to measure the amplitude, A, and the period, T, as shown in Figure 6.2. We find

$$A = 15.558 \text{ V and } T = 6.29\,\text{ms}$$

Next, we measure the coordinates of a point on the sinusoid in order to calculate the phase angle, θ, using Eq. (6.6). For example,

$$t_1 = 60.92\,\text{ms and } v_L(t_1) = 7.599 \text{ V} \Rightarrow \theta = 52.3°$$

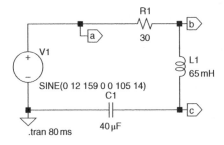

Figure 6.A.1 LTspice schematic from Figure 6.9.

Consequently, the steady-state response inductor in Figure 6.9 is

$$v_L(t) = 15.558\cos(1000\pi t + 52.3°)\,V$$

Next, to obtain a plot of the resistor voltage, *right*-click the label "V(b)−V(c)" in the space at the top of the plot to change "V(b)−V(c)" to "V(a)−V(b)." Do so, and then click "OK" to dismiss the dialog box. The plot will change to display V(a)−V(b), which is the resistor voltage.

Use the cursor to measure the amplitude, A, and the period, T, as shown in Figure 6.2. We find

$$A = 7.184\,V \text{ and } T = 6.29\,ms$$

Next, we measure the coordinates of a point on the sinusoid in order to calculate the phase angle, θ, using Eq. (6.6). For example

$$t_1 = 62.52\,ms \text{ and } v_R(t_1) = 3.660\,V \Rightarrow \theta = 38.0°$$

Consequently, the steady-state inductor current in Figure 6.9 is

$$v_R(t) = 7.184\cos(1000\pi t - 38.0°)V$$

To obtain a plot of the capacitor voltage, *right*-click the label "V(a)−V(b)" in the space at the top of the plot to change "V(b)−V(c)" to "V(c)." Do so, and then click "OK" to dismiss the dialog box. The plot will change to display V(c), which is the capacitor voltage.

Use the cursor to measure the amplitude, A, and the period, T, as shown in Figure 6.2. We find

$$A = 5.982\,V \text{ and } T = 6.29\,ms$$

Next, we measure the coordinates of a point on the sinusoid in order to calculate the phase angle, θ, using Eq. (6.6). For example

$$t_1 = 64.11\,ms \text{ and } v_C(t_1) = 3.138\,V \Rightarrow \theta = -127.6°$$

Consequently, the steady-state capacitor voltage in Figure 6.9 is

$$v_C(t) = 5.982\cos(1000\pi t - 127.6°)V$$

References

1 Svoboda, J.A. and Dorf, R.C. (2014). *Introduction to Electric Circuits*, 9e, 430. NJ: Wiley.

2 Svoboda, J.A. and Dorf, R.C. (2014). *Introduction to Electric Circuits*, 9e, 879–882. NJ: Wiley.

3 Svoboda, J.A. and Dorf, R.C. (2014). *Introduction to Electric Circuits*, 9e, 430–440. NJ: Wiley.

4 Svoboda, J.A. and Dorf, R.C. (2014). *Introduction to Electric Circuits*, 9e, 435. NJ: Wiley.

Chapter 7

Frequency Response

Suppose $v_s(t) = A\cos(\omega t + \theta)$ is the input to a linear, time-invariant circuit and $v_0(t) = B\cos(\omega t + \phi)$ is the steady-state response of that circuit. The network function, $\mathbf{H}(\omega)$, is the ratio of the response phasor to the input phasor

$$\mathbf{H}(\omega) = \frac{\mathbf{V}_0(\omega)}{\mathbf{V}_s(\omega)} = \frac{B\angle\phi}{A\angle\theta} = \frac{B}{A}\angle(\phi - \theta) \qquad (7.1)$$

The magnitude of the network function, $|\mathbf{H}(\omega)| = \dfrac{B}{A}$, is called the gain of the circuit, and the angle of the network, $\angle\mathbf{H}(\omega) = \phi - \theta$, is called the phase shift of the circuit. The gain and phase shift are both functions of the frequency, ω. These two functions comprise the frequency response of the circuit.

We can use LTspice to represent the frequency response of a circuit graphically by plotting the gain versus frequency and also plotting the phase shift versus frequency.

7.1 Frequency Response Plots

Example 7.1

The input to the circuit shown in Figure 7.1 is the voltage-source voltage $v_s(t)$. The response is the voltage, $v_0(t)$, across the $20\,\mathrm{k\Omega}$ resistor. The network function of this circuit is

$$\mathbf{H}(\omega) = \frac{\mathbf{V}_0(\omega)}{\mathbf{V}_s(\omega)} = -\frac{R_2}{R_1 + j\omega C R_1 R_2} = \frac{-4}{1 + j0.008\omega} \qquad (7.2)$$

Plot the frequency response of the circuit shown in Figure 7.1.

(Continued)

LTspice® for Linear Circuits, First Edition. James A. Svoboda.
© 2023 John Wiley & Sons, Inc. Published 2023 by John Wiley & Sons, Inc.

Example 7.1 (Continued)

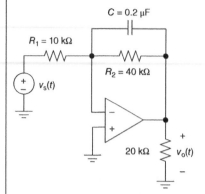

Figure 7.1 A circuit.

Step 1. Formulate a circuit analysis problem.

Determine the frequency response of the circuit shown in Figure 7.1.

Step 2. Describe the circuit using an LTspice schematic.

The circuit in Figure 7.1 contains an ideal op amp [1]. We can implement an ideal op amp in LTspice using the component named UniversalOpAmp2. The UniversalOpAmp2 has five terminals: the three terminals of the ideal op amp plus two additional terminals for power supplies. Suitable power supplies for the UniversalOpAmp2 used as an ideal op amp consist of 15-V voltage sources as shown in Figure 7.2.

To emphasize the correspondence between the circuit diagram in Figure 7.1 and the schematic, we will move the op amp power supplies to the left side of the schematic and use "Bi-Direct" terminals to connect the power supplies to the op amp as shown in Figure 7.2. (Click on the

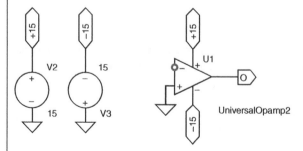

Figure 7.2 An LTspice schematic consisting of a single op amp and its power supplies.

Example 7.1   (Continued)

icon, ⎡ⓐ⎤, toolbar and select "Bi-Direct" from the dropdown menu named "Port-Type.")

Click on the New Schematic toolbar icon, ⎡⟲⎤, to start a new schematic. Place symbols representing the op amp and two 15-V voltage sources. Add the ground and some terminals and wire the circuit as shown in Figure 7.2.

The LTspice schematic shown in Figure 7.2 represents an ideal op amp that saturates [2] at ±15 V.

It is convenient to choose $\mathbf{V}_s(\omega) = 1\angle 0°$. Then, $\mathbf{H}(\omega) = \mathbf{V}_o(\omega)$ and the gain of the circuit is equal to $|\mathbf{V}_o(\omega)|$ and the phase shift of the circuit is equal to $\angle\mathbf{V}_o(\omega)$.

Place symbols to the schematic to represent the AC voltage source, resistors, and capacitor in the circuit diagram. Right-click the voltage source to specify that it is an AC voltage source with amplitude equal to 1 V and phase angle equal to 0°. Specify the capacitance and resistance values. Wire the circuit. Label a node. Figure 7.3 shows the LTspice schematic corresponding to the circuit diagram shown in Figure 7.1.

Step 3. Simulate the circuit using LTspice.

Select **Simulate/Edit Simulation Cmd** from the LTspice menu bar to pop up the "Edit Simulation" dialog box shown in Figure 7.4. Specify an "AC Analysis" with a Start Frequency of 1 Hz, a Stop Frequency of 1 kHz, and 100 points/decade.

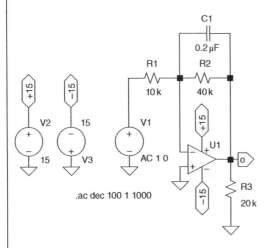

Figure 7.3   The circuit from Figure 7.1 described as an LTspice schematic.

(*Continued*)

Example 7.1   (Continued)

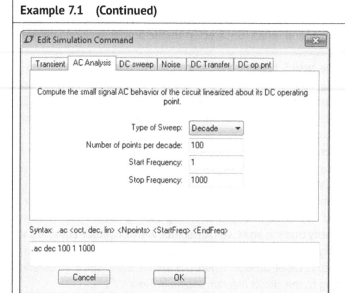

Figure 7.4   The simulation command dialog box.

(AC Analysis allows us to simulate a circuit over a range of frequencies. Some trial and error may be required to determine an appropriate range of frequencies. In this example, the network function has only one corner frequency, a pole at

$$\frac{1}{0.008} = 125 \, \text{rad/s} = 19.89 \, \text{Hz}$$

The "Start Frequency" of 1 Hz is somewhat smaller than the pole frequency, and the "End Frequency" of 1000 Hz is somewhat larger than the pole frequency.)

Four "Types of Sweep" are available: Octive, Decade, Linear, and List. This choice determines how the horizontal axis of the Frequency Response plot will be labeled. Chose "Decade" and click the "OK" button to generate a "spice directive" consisting of the command ".ac dec 100 1 1000." Position the spice directive on the schematic as desired.

Click on the Run LTspice toolbar icon, ⚡, to simulate the circuit.

Step 4. Display the results of the simulation.

After a successful AC Sweep simulation, LTspice displays a blank plot. Click on the output terminal, labeled "o" in the schematic, to obtain the plot shown in Figure 7.5.

Example 7.1 (Continued)

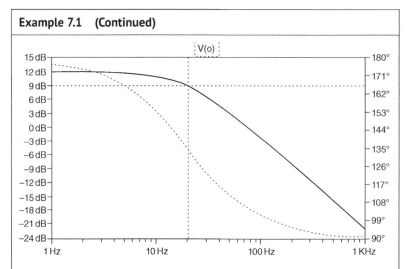

Figure 7.5 The gain and phase shift frequency response plots.

V(o) denotes the AC voltage at node o in the schematic shown in Figure 7.3. The solid line plot displays the magnitude of V(o), in dB, and the dotted line plot displays the angle of V(o). In Figure 7.5 a cursor is used to measure V(o) at the frequency 20.5 Hz. We see that at 20.5 Hz

$$V(o) = 8.895 \text{ dB} \angle 134.1°$$

The AC voltage V(o) is called $\mathbf{V}_o(\omega)$ in Eqs. (7.1) and (7.2). Because we have set $\mathbf{V}_s(\omega) = 1 \angle 0°$

$$H(\omega) = \frac{\mathbf{V}_o(\omega)}{\mathbf{V}_s(\omega)} = \frac{\mathbf{V}_o(\omega)}{1 \angle 0°} = \mathbf{V}_o(\omega) = V(o)$$

Consequently, the solid line plot in Figure 7.5 displays $\left| H(\omega) \right| = \left| H(2\pi f) \right|$ and the dotted line displays $\angle H(\omega) = \angle H(2\pi f)$.

Step 5. Verify that the simulation results are correct.

Let us check to see that simulation results agree with the network function of the circuit.

$$H(\omega) = \frac{\mathbf{V}_o(\omega)}{\mathbf{V}_s(\omega)} = -\frac{R_2}{R_1 + j\omega C R_1 R_2} = \frac{-4}{1 + j0.008\omega} = \frac{-4}{1 + j0.016\pi f}$$

When $f = 1.1415$ Hz, $H(\omega) = 3.993 \angle 176.7°$. Since $20 \log_{10} 3.993 = 12.027$ dB, the calculated gain agrees with the value labeled on the gain frequency response plot.

(Continued)

Example 7.1 **(Continued)**

Also, when f = 20.5 Hz, $H(\omega)$ = 2.7857∠134.14°. The phase angle agrees with the value labeled on the phase frequency response plot. Since 20 $\log_{10}$ 2.7857 = 8.895 dB, the calculated gain agrees with the value labeled on the gain frequency response plot.

The frequency response plots agree with the network function so the simulation results are correct.

Step 6. Report the answer to the circuit analysis problem.

Figure 7.5 displays the frequency response plots of the circuit shown in Figure 7.1.

Example 7.2

This example illustrates the use of global parameters to incorporate design equations into an LTspice circuit. It also illustrates that the LAPLACE part provided by LTspice makes it easy to verify that a frequency response corresponds to a specified transfer function.

The circuit shown in Figure 7.6 is a standard filter circuit called a Sallen–Key lowpass filter [3]. The transfer function of this circuit is

$$H(s) = \frac{V_o(s)}{V_i(s)} = \frac{k\omega_0^2}{s^2 + \dfrac{\omega_0}{Q}s + \omega_0^2} \tag{7.3}$$

where k is the DC gain, ω_0 is the corner frequency, and Q is the quality factor of the filter. The Sallen–Key lowpass filter can be designed to have specified values of ω_0 and Q using the following equations.

$$C_1 = C_2 = C = 0.1\ \mu F \tag{7.4}$$

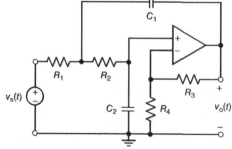

Figure 7.6 The Sallen–Key lowpass filter.

Example 7.2   (Continued)

$$R_1 = R_2 = R_4 = R = \frac{1}{C\,\omega_0} \qquad\qquad (7.5)$$

$$k = 3 - \frac{1}{Q} \qquad\qquad (7.6)$$

$$R_3 = (k-1)R \qquad\qquad (7.7)$$

Equations (7.4) through (7.7) comprise the design equations for the Sallen–Key lowpass filter. The quality factor and corner frequency determine the shape and position of the frequency response of the Sallen–Key lowpass filter. Given values of Q and ω_0 and a convenient value of the capacitance C, we calculate the values of R and R_4 using Eqs. (7.5) through (7.7).

Step 1. Formulate a circuit analysis problem.
   Exhibit the frequency response plot of a Sallen–Key lowpass filter having

$$\omega_0 = 1000^*\,\pi \text{ rad/s and } Q = 4$$

Step 2. Describe the circuit using an LTspice schematic.
   Click on the New Schematic toolbar icon, ⟨⟩, to start a new schematic. Implement the op amp in the Sallen–Key lowpass filter using the UniversalOpAmp2 and two 15-V voltage sources as shown in Figure 7.2.
   Add a SPICE directive to define parameters C, Q, ω_0, R, k, and R_k in accordance with Eqs. (7.3) through (7.7). Click on the SPICE directive icon **.op**, at the right end of the LTspice toolbar to open a dialog box. Enter

> .param $C = 0.1\,\mu\text{F}$, $Q = 4$, w0 = {1000* 2* 3.14159}

> .param R = {1/w0/C}, k = {3 − 1/Q}, Rk = {(k − 1)* R}

in the text box. Place the SPICE directive on the schematic as shown in Figure 7.7.
   It is convenient to choose $\mathbf{V}_s(\omega) = 1\angle 0°$. Then, $\mathbf{H}(\omega) = \mathbf{V}_o(\omega)$ and the gain of the circuit is equal to $|\mathbf{V}_o(\omega)|$ and the phase shift of the circuit is equal to $\angle\mathbf{V}_o(\omega)$.
   Place symbols to the schematic to represent the AC voltage source, resistors, and capacitor in the circuit diagram. Right-click the voltage source to specify that it is an AC voltage source with amplitude equal to 1 V and phase angle equal to 0°.

(Continued)

Example 7.2   (Continued)

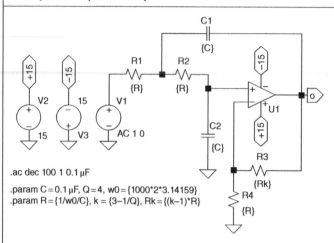

.ac dec 100 1 0.1 µF

.param C = 0.1 µF, Q = 4, w0 = {1000*2*3.14159}
.param R = {1/w0/C}, k = {3−1/Q}, Rk = {(k−1)*R}

Figure 7.7   The Sallen–Key lowpass filter circuit described as an LTspice schematic.

Specify the capacitance and resistance values. Wire the circuit. Label the output node. Figure 7.7 shows the LTspice schematic corresponding to the circuit diagram shown in Figure 7.6.

Step 3. Simulate the circuit using LTspice.

Select **Simulate/Edit Simulation Cmd** from the LTspice menu bar to pop up the "Edit Simulation" dialog box shown in Figure 7.4. Specify an "AC Analysis" with a Start Frequency of 1 Hz, a Stop Frequency of 1 kHz, and 100 points/decade.

Chose "Decade" as the type of sweep and click the "OK" button to generate a "spice directive" consisting of the command ".ac dec 100 1 10k." Position the spice directive on the schematic as desired.

Click on the Run LTspice toolbar icon, 🏃, to simulate the circuit.

Step 4. Display the results of the simulation.

After a successful AC Sweep simulation, LTspice displays a blank plot. Click on the output terminal, labeled "o" in the schematic, to obtain the plot shown in Figure 7.8.

Step 5. Verify that the simulation results are correct.

We will verify the simulation results by incorporating the transfer function into the schematic of the Sallen–Key lowpass filter and comparing the frequency response of the circuit to the frequency response of the transfer function (Figure 7.9).

Example 7.2 (Continued)

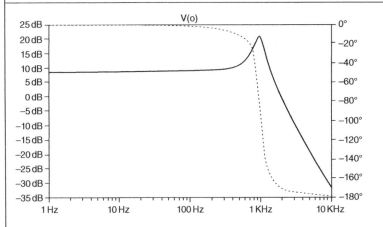

Figure 7.8 The frequency response plot of the Sallen–Key lowpass filter.

(a)

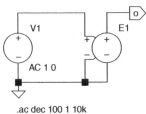

.ac dec 100 1 10k

(b)

Attribute	Value	Vis.
Prefix	E	
InstName	E1	✕
SpiceModel		
Value	Laplace=k*w0**2/(s**2+(w0/Q)*s+w0**2)	✕
Value2		

(c)

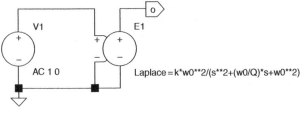

Laplace = k*w0**2/(s**2+(w0/Q)*s+w0**2)

.ac dec 100 1 0.1 μF

.param C = 0.1 μF, Q = 4, w0 = {1000*2*3.14159}
.param R = {1/w0/C}, k = {3–1/Q}, Rk = {(k–1)*R}

Figure 7.9 The schematic of the transfer function given in Eq. (7.3).

(Continued)

Example 7.2 (Continued)

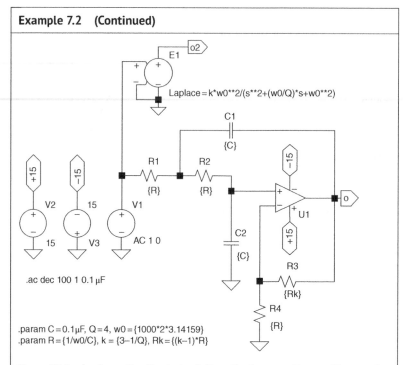

Figure 7.10 A schematic of both the Sallen–Key lowpass filter and its transfer function.

The schematic shown in Figure 7.10 contains both the Sallen–Key lowpass filter shown in Figure 7.7 and the VCVS shown in Figure 7.9c.

The schematic shown in Figure 7.9a consisting of an AC voltage source and a VCVS (Voltage Control Voltage Source).

The VCVS in Figure 7.9 implements Eq. (7.3) by including Eq. (7.3) in the schematic.

Right-click the VCVS (LTspice component E) to pop up a Component Attribute Editor dialog box. Edit the "Value" of the VCVS as shown in Figure 7.9b and then click "OK." Then define the required parameters by clicking on the directive icon .op to open a dialog box and enter

.param C = 0.1 μF, Q = 4, w0 = {1000* 2* 3.14159}

.paramR = {1/w0/C}, k = {3 − 1/Q}, Rk = {(k − 1)* R}

in the text box.

The revised schematic is shown in Figure 7.9c.

Example 7.2   (Continued)

Click on the Run LTspice toolbar icon, 🏃, to simulate the transfer function. After a successful AC Sweep simulation, LTspice displays a blank plot. Click on the output terminal, labeled "o" in the schematic, to obtain, again, the plot shown in Figure 7.8. The frequency response of the circuit is indeed the same as the frequency response of the transfer function VCVS implementing its transfer function.

Click on the Run LTspice toolbar icon, 🏃, to simulate the circuit and transfer function. After a successful AC Sweep simulation, LTspice displays a blank plot. Click on the output terminal, labeled "o," in the schematic to obtain the frequency response of the circuit, and click on the output terminal, labeled "o2," to obtain the frequency response of the transfer function. The plots are identical, overlapping exactly.

The design equations are correct and the simulation results are correct.

Step 6. Report the answer to the circuit analysis problem.

The Sallen–Key lowpass filter shown in Figure 7.7 has a corner frequency, ω_0, at 1000 Hz and a Q equal to 4. Figure 7.8 shows the gain frequency response plot of the Sallen–Key lowpass filter.

References

1 Svoboda, J.A. and Dorf, R.C. (2014). *Introduction to Electric Circuits*, 9e, 219. NJ: Wiley.

2 Svoboda, J.A. and Dorf, R.C. (2014). *Introduction to Electric Circuits*, 9e, 245–247. NJ: Wiley.

3 Svoboda, J.A. and Dorf, R.C. (2014). *Introduction to Electric Circuits*, 9e, 810–811. NJ: Wiley.

Index

LTspice® for Linear Circuits, First Edition. James A. Svoboda.
© 2023 John Wiley & Sons, Inc. Published 2023 by John Wiley & Sons, Inc.

Printed and bound by CPI Group (UK) Ltd, Croydon, CR0 4YY

27/10/2024

14580673-0005